The Oracle

Principles of Creative Sciences

The most complete and authoritative course on Creativity and creative success; a core subject in the tertiary institutions and the professional creative scientist.

Peter Matthews - Akukalia

World's 1st Creative Scientist

AuthorHouse™ UK Ltd.
500 Avebury Boulevard
Central Milton Keynes, MK9 2BE
www.authorhouse.co.uk
Phone: 08001974150

Published by AuthorHouse 2/6/2012

ISBN: 978-1-4678-8312-2 (sc)
ISBN: 978-1-4678-8313-9 (e)

A GREENDOVE SERIES BOOK.
The acronym: P.M.AKU is the reference points of all creative laws, principles, philosophies, formulae, and diagrams; stated or made in this work. Also of interest is the acronym: pmaku (pronounced as ma-ku, silent p sound, such as is used in Pmakumatics or Pmakumeter guage etc. They are both acronymic of the full name of the author of this work: Peter Matthews - Akukalia.

This work is not complete without the Pmakumeter guage.

Front page diagram:	*The Shining Stars of Ideas and Inspiration, flanked by a medal of merit.*
Section 1 diagram:	*Podium of Creative Success.*
Section 2 diagram:	*A book and candle as the shining lights of knowledge.*
Section 3 diagram:	*Bundle of cash representing creative entrepreneurial Financial success.*

BRIEF BIOGRAPHY OF AUTHOR

Peter Matthews Akukalia (A.k.a P.M.AKU) is an aviator, educator, poet, playwright, novelist, compiler, private researcher, founder, tutor, educational philosopher, a conference maker, a creator, a creative scientist, a disciple of entrepreneurship, a facilitator, a committed advocate of functional practical education as well as an expert in youths and, skills resourcefulness and development. Though obtained a training in Aviation, he discovered his calling in Education, creativity and strategy and has since lived by it. He is an experienced teacher, a former vice-principal, a former tutor, a former Lecturer in business law and communication skills at Senate Tutors, Lagos {ICAN INSTITUTE}, a private researcher on functional education and youth development, as well as a career analyst, for secondary schools and adults as well, with resounding results.

As a creative Scientist, he has won himself, the 1st Charted sage Ch.S. A title in the field he authors as its Pioneer and founder. As a creator, he created the Educational Development Master Plan (EDMP) presently in use, he also created the special genres of Literature known as the Propoplay and Nidrapoe, with books to its effect. He created the popular students Career Analysis program (SCAP). As an author, he has many works to his credit including poetry, novel, Biography, Play, Propoplay, and project works. His other experiences include working as a sales representative in a private chemical company, Ondo State and on board foreign ships and vessels as an engine room Assistant.

He is a Christian. Born on 6th July, 1978, in Lagos. He had his primary education at Jack N'Jill Children School, presently located in Ilasamaja area; then his junior secondary at Ansar-ud-Deen Comprehensive High School Okota, and completed it at Onitsha High School, Onitsha, Anambra State. He hails from Nkpologwu, Nsukka Uzo-uwani Local Government Area in Enugu State. Nigeria.

He is a true Patriot. He also widely traveled. Presently, he towers as the President/CEO; UTMOST PEAK RESOURCES, FOUNDER and

PRESIDENT - INSTITUTE OF CREATIVITY & MANAGEMENT STUDIES. He is a force to be reckoned with in classics and creative sciences. His philosophies are Resourcefulness, Diligence and Reward.

He is variously esteemed as African Czar; for his dogged belief in the power of Education as a bedrock of any society; as African Shakespeare for his unique ability to write in the Shakesparean style and then as the sage: for his creative prowess. His works speak for him. He is a great gem. By this complete work on the course of creativity, he has become the world's 1st creative scientist in the history of mankind. A hundred steps for intellectuality; a million flights for civilization. It is simply amazing.

Peter Matthews - Akukalia

ACKNOWLEDGMENT

Brief as it may be, my gratitude shall be expressed to the Almighty God who inspires all works for good. Then I appreciate the staff and workers who directly or indirectly contributed to the successful completion of this work.

Thank you and God bless you.

Specially Dedicated
To:

Nigeria, and all Nigerians
Home and in diaspora
On the most deserved celebrations
Of our 50th Independence Anniversary
Our Golden Jubilee:
Remember, I took the lead This time
But keep this.

And also to:

Late Diana Matthews; my granny, for teaching me the practicals;
My Uncles; Kwesi and the likes who built my confidence
Whom I admire in pictures, Ameachi who showed me home and gave me my place, Victor, who sharpened my mind
Paul, who gave me a heart; Chuka, for his commanding height,
Tony for his intellectual might, how he stupefies those with no foresight the last of the blacks after my dad; My dad, though late, for those materials on education was his enduring legacy; Aunty Rita who inspired me to aspire she loved my shoulders and let me taste power once, preaching education for the girl child;
Mrs. Jojolola Odujoko; my proprietress
Mrs. Ijeoma Elobuike, my headmistress; the two who refused to give up on me;
Late Mr. Benson, who taught me to speak Good English.
Mr. Bordoh; who in strict discipline, taught me Mathematics
Brother Austin, for the private classes in Geography;
My mum, who endured all the troubles, and a funny siblings;
Madam Ijeoma; my friend, who works and smiles helping to endure the bites of works;
Police, Israel, Ufoma, Palee, Goddy and the others upon the vessels who still urge me to keep the tussles;
Cosmos who assisted me without excuse, and his love for reading;

Jecintha who showed kindness even to her own bruise until Heaven rained down kisses,
My wife and kids, none the least; for theirs is full of gists;
Then to my friends so concerned, Chuks, Godwin Eze, and the reader,
To Nigerians and the education sector even the teachers; whom Heaven hands over the Registers of earth's keepers;
Then the youths, students and pupils who will one day save the honours;
Then to myself, God's instrument for the works,
Above all to God my greatest inspiration and teacher for giving me these lectures.

INTRODUCTION

Your excellency my dear student. There is a world greater than any thing else. It is the world of creators. Their task is creativity. The issues bothering on the wheels of progress in our collective humanity are nothing new to the common populace. Over and over again, history has proved that those who are most celebrated for outstanding contributions to humanity were creators. It is either they were honoured for their literary and intellectual works or their scientific inventions. These people by personal and careful study, have built what I refer to as a "Silent Empire". Nations and people have profited and shall continue to profit by their works.

Creativity is the last hope of mankind and his first requirement upon the very earth on which he threads. I predict that one day nations will begin to assess the strength of their leaders by their own personal works of creativity as well as its direct impact on society; not necessarily wealth. I also foresee in my third eye the creation of a special sector of creativity with a ministry such as a Ministry of Creativity; managing its affairs. I foresee schools adding the subject titled: Creativity and institutions with the name school or college of creativity where people who have discovered themselves beyond mere speech but with sample works would be trained formerly to lead the "silent Empire".

This book has been written in support of this vision. It is strictly the originality from my mind. Based on my various experiences I shall not be pleasing my creator nor helping humanity if I do not script the secret keys to the personal success fish I have been frying; especially in my country, if I do not place on; obliged, beyond the mere systems in place, to show the given grace of leading the pace in the issue of not the general success, for even a stupid person can make money ; but creative success that which comes from the within of a person; is original, has value, adds beauty and praises God- the source of all inspirations.

Creative success involves a process. For this process to work for both the science or arts oriented creator certain principles have been broken down to readable laws, deduced formulae, and general illustrative diagrams enough to form a course. I hope to hear from you. See you at the top; the point of creative success. Whether the professor or the inventor, the critic or the educator, the simplest of teachers, the certified or the experienced or even the adherent of both; we do well to acknowledge the fact that the world in which we live is a rolling ball. The simple factor behind this dynamism is knowledge - unrestrictive knowledge. We can only command our tomorrow, the future by the extent of knowledge we grapple today.

I have written and compiled with a certain sense of appreciation as the experts do, in a field of research born of my own initiative in order to initiate the mind of the reader with the strength and enormous possibilities behind this encompassing field called education; what my research findings of many years has to offer. This might seem strange, but a careful consideration of the principles stated therein will prove its merit for innovations and no contrary sides.

When a certain knowledge is bottled to brim within the human mind, it heats up until it can no longer stay. As a boiling water, it either steams out gently causing warm effects or bursts with force breaking the bottle. I have allowed for the first. As an intelligent, rational and realistic person, I think it the best of interests to pen down these new findings of mine; principles, ideas or what I feel may find its way one day into the fundamentals of our basic studies. These stated simply as laws and deductions I hope shall aid our reasoning faculty to sincerely evaluate our present system to yield better results for the generations yet unborn.

Scholarships? They are additions to place your own future in your own hands. They have existed for donkey years. This is the first time, the average child on the street will have access to it. It is only an extra benefit, just like we spice up our cake after baking.

This book is made by popular demander and compelling years. I believe it has come at the right time. Education belongs to all.

Perhaps, persons of notable worth who wish to understand the source of these could see my letter on the section Creative Education then others become interesting.

This book is a gift to humanity certainly so.
Happy reading, Happy reasoning.
- Peter Matthews Akukalia

General Aims / Objectives of the PCS.

1. To consolidate on the need for economic empowerment through personal development.
2. To train the mind creatively.
3. To breed a new generation of creative scientists as a field of study and career pursuit.
4. To train every student to become a strategist in order to become self-employed creativity.
5. To resuscitate the various sectors by training creative scientists and strategists.
6. To aid the establishment of the Ministry of Creativity to recognize creativity in all its ramifications.
7. To promote the ideals of the Institute of Chartered Creative Scientists as well as other professional organisations set up for this purpose.
8. To achieve the objectives of the Millenium Development Goals (MDG) and reach for a greater civilization and a thriving economy beyond it, ushering in the transition from the present information age to the creative age, or moreso, the info-time age.
9. To aid self-employment, while fulfilling the understanding of human behavioural science and development.
10. To fulfill the need for a comprehensive course on creativity suitable for adoption to the Education System, as a General Study Course (G.S) or as a core subject for universities and Tertiary Institutions thereby promoting a Functional educational System.

SEVEN QUALIFYING CHARACTERISTICS OF A SCIENTIST.

1. Source of study based on discovery.
2. Direct personal involvement comprising experiences, researches and results
3. Applicable philosophies, comprising theories, hypothesis, logical explanations, and briefs.
4. Derivable laws e.g Newton's 1st, 2nd, 3rd laws of motion Lechatelier's law of equilibrium etc.
5. Developed calculations: simple and complex.
6. Developed instruments and parameters for measurements (often calibrated and Confirmatory).
7. Judgement / Analysis of impact on humanity and the economy; divisible into:

a. Guiding regulations (legal or moral)
b. Cause
C. Effect
d. Problems
E. Solutions

Please note that when primitive problems are solved, then it leads to civilization. Creativity confirms dynamism and inevitability.

Sections: Intro

The Oracle: Principle of Creative Sciences is divided into three sections:
Section 1: Human Creativity
Section 2: Creative Education
Section 3: Creative Entrepreneurship

And

Two parts:
1. Creative Doctrines
2. Pmakumatics.

Further studies would explain these better.
STRATEGIC OPERATIONS STATEMENTS (STOPS).

Stops 1:

If you knew perhaps what the creators know;
Who knows you might start creating also;
For impact and significant living.
- Peter Matthews - Akukalia

Stops 2:

I have termed creativity - The silent Empire
because it is a fire burning within the person
making him the potential king of his life and
helping to lead others in the same arena
then defining his true place in life's own
scheme of things.

- Peter Matthews - Akukalia.

Stops 3.

Creativity has no apology nor neutrality,
It's either burning for greatness, or
Destroying against great hurts;
It's either discovered for usefulness,
Or denied against our own benefits.

Stops 4:

Creativity can be defined as the ability to produce original ideas or things. No one seems to know why one person may be more creative than another........

- (Drewry, O' Connor)

Stops 5:

Not he is great who can alter matter,
but he who can alter my state of mind.....
now at last ripe., and inviting nations to
the harvest. The great man makes the
great thing

- Ralph Waldo Emerson
(August 31, 1837).

Stops 6:

But now you are great, my students; now no other time,
but now my fellow country men because by this work,
your mind is altered for good; for greatness,
because now you know why.

- Peter Matthew S Akukalia.

Stops 7.

It all began with the Agricultural era, then the industrial ; then the computer era: then the jet age; and then the information age; but now it is the info-time age - the new era of success not determined by age but the proper application of information and skills massively with a success production rate at a short time; this work officially welcomes you to the info -time age; the creative age ; starting from now!.

We are not only ushering in the creative Age; we are also taking the lead at creating it. We are prepared; we are ready and we hope you are.

- Peter Matthews - Akukalia

.....There's no mystery to this, it's all about discovering and developing you.

- Peter Matthews - Akukalia.

TABLE OF CONTENTS

SECTION 3: CREATIVE ENTREPRENEURSHIP (THE BUSINESS ASPECT)

SECTION 1:

Human Creativity

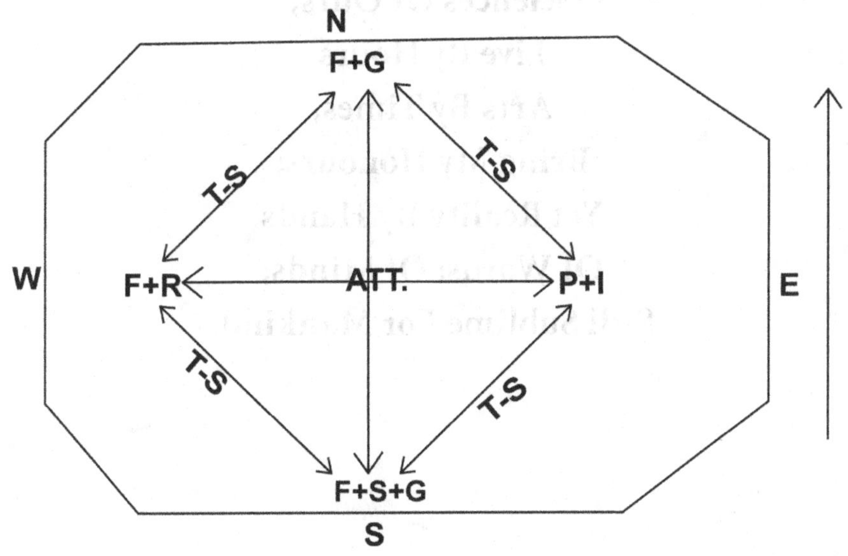

Sciences Of Ours,
Live By Hours
Arts By Times,
Bring My Honours;
Yet Reality By Hands
Of Words; Of Minds;
Still Sublime For Mankind.

THE CONCEPT OF CREATIVITY

AIMS / OBJECTIVES

This chapter aims to help the student understand the following:

(I) The proper definition of Creative Sciences

(Ii) The relationships of Creativity in respect to its existence as either a fact or a mere myth;

(Iii) The purposes of studying Creative Sciences as well as its benefits;

(Iv) Certain examples as drawn from the historical perspectives;

(V) The classes of the oracle which relate to:

(a) Creative Doctrines

(b) Pmakumatics

(Vi) The differences between the study of Pmakumatics and the Conventional Mathematics as well as their similarities.

CHAPTER 1

THE CONCEPT OF CREATIVITY

WHAT IS CREATIVE SCIENCES?

The term "Creative Sciences" is coined from two words "Creative" and Sciences" defined in the following ways:

The Collins Gem English Dictionary (British) defines them as:

* Creativity - Imaginative or Inventive
* Creator; One who creates
* Science: Systematic study and knowledge of Natural or physical phenomena.
* Scientist - Person who studies or practices a Science. The American Heritage

  Dictionary defines as thus::
* Create - To cause to exist:
* Creation: The act of creativity; a product of invention or imagination; the world

  and all things in it.
* Creative: Characterized by originality; imaginative.
* Creator: One that creates.

* Science: The observation, identification, description, experimental investigation, and theoretical explanation of phenoma; methodological activity, discipline or study; an activity regarded as requiring study and method; knowledge gained through experience (obtained from the latin word; Scientia meaning knowledge).

Scientist: A person having expert knowledge of one or more sciences.

From the above, we may define Creative Science as a specialised knowledge based on systematic study of the process of inventiveness; or bringing to full use the knowledge of developing an original work from original ideas.

A person who has an expert knowledge of creativity and creative sciences, is called a Creative Scientist; while he becomes a creator, only after he has made an original work and caused it to come into existence for human use and development.

CREATIVITY - FACT OR MYTH?

Over the years, people of various nations and societies have considered the issues of creativity, endowment, potentials and talents. It is proved that the creator of the Universe made man in His special image and likeness - also in reference to certain qualities and behaviour endowed man to a certain capacity, to develop his uniqueness and thereby bringing happiness to himself and the world around him. This is called the principle of creationism. This is defined as the position that the biblical account or more so, the religious beliefs of creation is literally true. The person who studies this is known as a creationist.

According to the book: America is " creativity can be defined as the ability to produce original ideas or things. No one seems to know why one person may be more creative than another. No one seems to know why creative activity seems to occur at one place over a period of time. Some people say it happens when people recognize a need for something".

"Facts are pieces of information based on impartial reality. Opinions are judgements or view points about a particular subject; but a myth is a story or narrative told, with a sense of questionability. We are not sure whether the story is true or false, except it is supported by great historical evidences". Creativity is a fact because it has provable circumstances, situations and works. Here in the latter study, we shall endeavour to demystify this study called creativity.

1:3 PURPOSE / BENEFITS OF STUDYING CREATIVE SCIENCES.

1. TO DISCOVER ONE'S UNIQUENESS

The principle of creationism, as well as our daily experiences proves that man as an individual is special and unique. This is proved by our various works at different dimensions, places and time.

2. ADDITION OF VALUE AND BEAUTY

Creativity promotes better values and beautifies a creation. The essence of creativity is to promote originality and help it work. Yet, in order for it to make the necessary impact and appeal to the eyes, it must be beautified, either by design or colour. Branding is the name given to a creation which must be valued by the mind and its true strength for humanity.

3. SELF EMPOWERMENT.

A creator has the benefit of being self-employed if he can find the right use, and market for his creation. This ensures less liability upon the society chest and government. He is responsible and accountable for his works. He also enjoys personal freedom.

4. SPIRITS OF DEMOCRACY AND PERFECTION

A Creator enjoys certain rights, as long as they are not destructive. These include:

I. Spirit of democracy: this involves his choices of what to create; for whom, how and his points of sale, its message and means.

Ii. Spirit of perfection: As long as he is alive, he has no limit to the extent of his creativity. He can grow by his works and earn the prestige depending on the quality of his works.

5. SENSE OF BELONGING:

Every one wants to be known for something, to become popular, to become useful. This is called the sense of belonging. Creativity ensures that the creator has a personal esteem and is important. It breeds confidence and not conceit.

6. VALUE OF PERSONS / LIVES:

Since everyone is endowed with potentials, it also proves that everyone has a certain solution to everyone's specific problem. Thereby, it is a conclusive fact that creativity must value everyone; every life is important when judged by impact. The citizens of a nation make that nation. The greatness of a nation surely depends upon the extent of respect and attention payed to creativity and must be based upon the principle." Everyone for himself; everyone by each other".

7. STANDARD OF LIVING.

There is often a good standard of living in countries and so cities where creativity is the basis of survival. Everyone finds the space where he belongs and will be know. He therefore earns from there.

8. COMPETITION:

Healthy competition is necessary in every society. When people are creative and competitive, it leads to innovations (better ways of doing the same thing or perfecting the same operations); thereby leading to low costs of purchase, better standards; efficiency and overall development of the economy.

9. INDUSTRY:

The fruit of industry in a country or society is a motivating factor towards true patriotism. It makes for hard work, perseverance, tolerance, and nationalism. The creator promotes industry by his works and point the way to proper culture and civilization.

10. ACCESS AND RECORDS:

There is no limit to a person who is creative. It is not limited to age, sex, race, qualifications, class or position. The extent of a creator's greatness surely depends on none of these, but on the extent of his works. Everyone has access to it. The creator has access to excellence if he can work at it as well as keep records in order to guide future generations.

1:4 HISTORICAL PERSPECTIVES

Here, we shall discuss briefly few creators and how their works affected mankind.

LITERATURE

1. WOLE SOYINKA:
Wole Soyinka was born on July 13,1934, in Abeokuta, Nigeria. He is a dedicated opponent of political oppression and corruption which are often the underlying themes of works. Soyinka writes plays novels, and poetry, as well as autobiographies, such as " Ake: The years of childhood" and Isara, a voyage around "Essay". In 1986, Soyinka became the first African writer to be awarded the Nobel prize for literature.

2. RALPH WALDO EMERSON:
A new philosophy or set of ideas, was developed largely by Ralph Waldo Emerson of Massachusetts, called transcendentalism, it is also called the spirit of perfection. Emerson believed that people could go beyond their limitations and perfect themselves and their society. Transcendentalists thought that every person was worthwhile, and that people had the ability to guide their own lives. Their philosophy was hopeful and encouraged the idea of progress. Although the transcendentalists did not join the reform movement, their ideas led to reform. In the years between 120 and 1850, a number of people in the united states recognized the need for a uniquely American Literature. One of the literary figures of the time was Ralph Waldo Emerson. The selection that follows is from Emersons "Nature", published in 1836.

"So shall we come to look at the world with new eyes. It shall answer the endless enquiry of the intellect what is truth?... What is good?...Then shall come to pass what my poet said; Nature is not fixed but fluid. Spirit alters, moulds, makes it every spirit builds itself a house and beyond its house a world and beyond its world a heaven. Know then that the world exists for you...... Build therefore your own world. As.......... you conform your life to the pure idea in your mind....... a revolution in thing will attend the influx of the spirit and the advancing spirit creates its ornaments along its path, and carry with it the beauty it visits and the song which enchants it, it shall draw beautiful faces; warm hearts, wise discourse and heroic acts, around its way, until evil is no more seen...."

On August 31, 1837, Emerson delivered" the American Scholar" speech at Harvard college. At that time, said the following.

"Not he is great who can alter matter, but he who can alter my state of mind. They are the kings of the world who give the color of their present thought to all nature and all art, and persuade men by the cheerful serenity of their carrying the matter, that this thing which they do is the apple which the ages have desired to pluck, now at last ripe, and inviting nations to the harvest. The great man makes the great things."

B. INVENTIONS AND INDUSTRY

New inventions were as important to the growth of big business. Some inventors in the united states worked to improve the production of goods. In the 1850's, Henry Bessemer of England, William Kelly of the Unites States separately discovered a new way to make steel from iron ore. This process made it possible to make more steel at less cost, and steel industry grew rapidly.

Other inventors worked to improve communications. In 1867, Christopher Sholes developed a typewriter. By 1873, the Remington Company was making typewriters on a large scale. In 1876, the telephone, invented by Alexander Graham Bell, was introduced in Philadelphia. By the 1890's, the American telephone and Telegraph company had installed nearly 500,000 telephones in America's business and homes. Thomas Edison made some major improvements on the telegraph. Soon, telegraph lines reached every part of the country and were used by business to carry out wide spread operations. Thomas Edison was called " The Wizard of Menlo Park". In his laboratory in Menlo park, New Jersey, he experimented with new ways to improve industry.

Thomas Edison also made other contributions. In 1879, he introduced the first practical electric bulb. A short time later, he invented a dynamo to generate electricity. Then in 1882, he set up the first central power plant in New York City. Electricity soon became an important source of power for homes, offices and industry.

Inventions helped the growth of industry. The need for new ways to do things produced a large number of inventions in the 1800's. In 1800, there were 309 patent - licences to make, use, or sell new inventions - registered with the United States Patent office. By 1860, there were 28,000 new patents.

Eli Whitney was responsible for several important discoveries. In 1793, Whitney developed the cotton gin (short for " engine"), which removed the seeds from cotton. Using this machine, workers, were able to prepare more cotton for shipment to textile mills in a shorter time. Whitney also developed a new system of interchangeable parts in the making of fire arms. These were parts that were exactly alike. If one part of a gun was damaged, another part of the same kind could replace it. This discovery made possible mass production. Other inventions were made according to the need and economic sector, and has continued so up till this day. (Adapted from America is by Drewry and O'connor).

There is need for the student and creative scientist to understand the basicity of this history. This is simply the fact that citizens, people, rose from grass to grass, from shabbiness to civilization; by creating, by using their heads and focus, having the right mind set and working to bring efficiency into their economy. A creator takes responsibility, initiative and is self -reliant.

When he turns his works to big business, then he becomes a creative entrepreneur. The rights of a citizen is clearly understood and applicable when such one is useful.

We shall study the basic principles of talents and potentials in latter chapters. The essence of this chapter is to help us appreciate the special abilities, given to us as individuals and collectively appreciate this. Let us consider the following statement:

1. Before we can set out on the road to success, we have to know where we are going, and before we can know that we must determine where we have been in the past.
   John F. Kennedy.
2. You must obey this, now, for a law - that he that will not work shall not eat.
   John Smith, English Colonizer, to sellers at James to win (1608)
3. Always do right. This will gratify some people, and astonish the rest.
   - Mark Twain (Samuel Langhorne Clemens), from a speech to the Young People's Society, Greenpoint Presbyterian Church, Brooklyn, N.Y. (February 16, 1901).
4. No race can prosper till it learns that there is as much dignity in tilling a field as in writings a poem.
   - Booker T. Washington, from Up from Slavery (1913).
5. Genius is one percent inspiration and ninety-nine percent perspiration.
   - Thomas Alva Edison, from life (1923)
6. Intellectually I know that America is no better than any other country; emotionally I know she is better than every other country.
   - Sinclair Lewis, American writer (1885-1951) from a speech.
7. It's what we learn after we think we know it all that counts.
   - Frank Hubbard, American humorist (1868- 1930), from back country Folks.
8. The half- baked ideas of people are better than the ideas of half- baked people.
   Professor William B. Shockley. American inventor and eugenics theorist, from Esquire (1973)
10. Scientists and artists perpetually live on the edge of mystery, being always surrounded by it.
   - J. Robert Oppenheimer, American Physicist, quoted in contemporary Architects (1980).

   Though this study is collectively known as the principles of creative sciences, a distinct separation may be discovered, observed and noticed. This division is classified into two classes.

CLASSES OF THE ORACLE: P cs

1. Creative Doctrines:

This is the section or class which discusses the various philosophies, principles, laws as well as the personal, experiences present in its discuss. These, together, are classified as Creative Doctrines. They are often named after the author and creator of the field of creative sciences as P.M. Aku's principle of....etc. This qualifies him to become the world's 1st creative philosopher ever produced.

2. PMAKUMATICS

The second aspect of this study consist of the various diagrammatic representations e.g.. the learning constituents Triangle (LCT); triangle of knowledge; cycle of professions etc; the calculative deductions e.g.. the level of curiosity, level of Endowment, level of interest, the statistics of creativity etc; then the various formulae developed on its own merit as well as the instruments (cash Pmakumeter Guage) developed for practical purposes. These collectively form the aspect of creative sciences known as Pmakumatics: as named after the author. This forms the authority for the training of creative scientists as strategists for future use in research works and human development. It may also be termed Creative Mathematics. The results are often subject to relativity of the individual and based on his records of personal interests database, psychoanalysis test or personal attitude chart etc.

It measures the uniqueness and average contribution of a citizen to his community without any excuse. It studies the measurement, properties, and relationships of quantities, using numbers and symbols of creative principles and creativity. If a creative scientist decides to major or specialize in this aspect, he would be practically referred to as a Pmakumatician (note it is pronounced with the P.sound silent i.e makue- ma-tish 'en). While the conventional mathematics limits perfection to 100%, Pmakumatics has an upper limit of 1000% IE (Infinitum Endowment) known as the Peak of Aspiration (POA).

The conventional mathematics is based on physical matter and so is subject to further adjustments, the Pmakumatics bases its study on the interest and characteristics as exhibited from the choices from the mind expressed on paper. Mathematics expresses failure and passes, Pmakumatics expresses level of intelligence and potentials which are subject to further development. Listed below are the differences and similarities between both forms of calculations presented in a tabular form.

Differences between Pmakumatics and Mathematics.

PMAKUMATICS	MATHEMATICS.
1. Has minimum limit of Zero (0) and upper limit of 1000% IE	Has minimum limit of zero (0) and upper limit of 100%
2. Calculates inherent personalities and interests of mind.	Calculates physical matter.
3. Has a perfection point called Peak of Aspiration	Has a perfection point called Excellent.
4. Works on the principles of creative endowments as expressed by person's choice and records.	Works on one's instant ability to solve certain questions.
5. Requires higher Intelligence to brilliance ratio.	Requires higher Brilliance to intelligence Ratio.
6. Respects every person's dignity.	Respects the selective ability of few with mathematical acumen.
7. All encompassing necessary for every field of study.	Necessary for science based scholars.
8. Suitable for our present state of Civilization and practical life.	Dates from old; necessary as a Study foundation.
9. Uses the L T / LT to measure cc/C² Quantities and CASH.	Uses the meter rule types to measure Quantities.
10. Measures the energy level, TTL, level of curiosity and interest and mind before creativity is established.	Measures physical quantities of matter creatively produced by the mind.
11. Has a Nigerian (African Origin) - single study	Has a Roman / Greek origin-group Studies.

Similarities Between Pmakumatics and Mathematics

1. Involves calculations
2. Works by formulae
3. Developed from systematic studies
4. Both respect the results of creative potentials
   Since they both lead to results of creativity.
5. They both use symbols and calibrated
   Instruments to reflect degree of accuracy.

SUMMARY

1. The term "creative sciences is defined as a specialized knowledge based on systematic study of the process of inventiveness or the specialized knowledge of developing an original work from original ideas.
2. The myth of creativity, once a mystery has been demystified through a comprehensive research creating it as a fact, a reality through the principles of creationism and other factors as observed from history.
3. The combined purposes and benefits of studying creativity include:
   A. To discover one's uniqueness;
   B. Addition of value and beauty;
   C. Self employment ;
   D. It promotes spirit of democracy and perfection;
   E. To gives a sense of belonging;
   F. It adds value to lives and persons;
   G. Enhances standard of living;
   I. Promotes healthy economic competition
   J. Promote industry in a nation leading to further innovations and inventions
   K. It creates access for every one and proper accountability through record keeping.
4. The classes of the oracle principles of creative sciences include:
   A. Creative doctrines: which discuss the various theoretical aspects of creativity such as personal experiences; philosophies, laws as propounded in this study;

B. Pmakumatics: which proves the various calculation and calculative deduction based on the study of creative sciences.

5. The various differences between Pmakumatics and mathematics arise from their distinguishable characteristics; while independent of the similarities between them.

CHAPTER 2

THE SPIRITUAL ESSENCE OF CREATIVITY

Aims/ Objectives:

1. To prove that the study of creativity does not jeopardize true religion but discusses it as the very essence of universal existence;
2. To help the student appreciate himself as a product of creativity.
3. To help the student understand the guiding philosophies behind certain stated portions of the scriptures in relation to creativity.
4. To prove that everything seen and unseen were made by the power of a great intelligence known a the creator or the Almighty God.
5. To apply fact behind the school of thought known as creationism as a support factor for man to continue to create positively.
6. To relate the creative facts of the bible as a reference point for every other religious book to support creativity.
7. Every religion is personal, but service to God through love and creativity is universally accepted.

The Spiritual Essence of Creativity

In dealing with a concept as crucial as creativity, it is highly necessary to consider its very beginning. This beginning is founded on the soils of the spiritual essence behind for where science disproves religion then there is a missing link. "Man was made", is a scripturall essence which disproves Darwinism theory of evolution. If so, then an intelligence brought man into the earth in which he found himself.

The Bible, used here as a point of reference exposes us greater to the study of creativity. Herein, we shall see also the purpose behind creativity.

Genesis 1:1 " In the beginning God created the heaven and the earth"

Guiding Philosophies:

I. There is no creativity without intelligence and origin.

Ii. When the work of creativity is over, it becomes a historical perspective and reference

Genesis 1:2: "And the earth was without form, and void and darkness was upon the face of the deep. And the spirit of God moved upon the face of the waters".

Guiding Philosophies

Ii.. Creativity occurs most where, and when a need arises; i.e creating occurs at the recognition of a need or problem.

Ii.. If the attitude of a will loses focus on the solution to a need, then it is likely that no creativity occurs.

Iii.. Problems must be studied from their deep causes, and deep rooted foundations sought for them.

Iv.. The instance of a problem also demands an instance for an action. Reading vs. 3- 31; we may deduce the following philosophies.

(1). Though a problem seems simple on the surface it may occur as a package of many others.

(2). When the segment of a problem is solved, the tendency to surface in another way is possible.

(3). Creativity involves finding a useful purpose where there seems to be none; and blending in correctly.

(4). "And God called..." Implies that branding is a useful point of creativity".

(5). As problems multiply and cause dissatisfaction, so do solutions multiply and bring happiness.

(6). "Every problem has an applicable solution. Vs 27 says: So God created man in His own image, in the image of God created he him, male and female created He them".

(7). If God created man in His image and likeness, and more so, everyone at that; then it is highly provable that man (everyone) is endowed with a certain spirit of potential in him. These potential are called talents.

(8). ...implies that creativity and discovery for human profitable use, respect and dignity in proper service to the true God, has won its approval from God ever since the days of creation. Vs 28(b)..."And God said unto them, Be fruitful, and multiply, and replenish the earth, and subdue it; and have dominion over the fish of the sea, and over the fowl of the air, and over every living thing that moveth upon the earth".

(9). The Spiritual essence of man"s existence is worship to God, his physical essence is creativity and service.

(10). The greatest resources of creativity are deposited in the mind; the others are of the earth. Vs 31:" And God saw everything that He had made, and behold, it was very good".

(11). God acknowledged the word "good", because He had set the foundations of the Earth, ready for use by man in order to make it better for his own living.

This process involved work. A further study in chapter 2 proves that God rested from the creation of the earth and then made man in order to let man take part in the joys of creating. Interestingly, man named those creatures which were not named by God. God gave a general name e.g fowls of the air; man named them specifically e.g eagle, kennery etc. Creativity brings multiplicity, growth, intelligence, discipline, resourcefulness and purpose.

The strength and might of God's abundant creative potential may be studied in God's answer to in Job 38,39. This discussion refers to the elements, as well as living creatures. God is diverse in ability and proves man a subject of developmental process to be perfected through faith and good works.

Psalm 82: 6 reads "I have said, ye are gods, and al l of you are children of the Most High".

We should not allow ourselves to be controlled by the evil gods of economic hardship, fear or terror in the fulfilment of our original purpose.

Painful as it were, we have given ourselves often times to wrong information, attitudes and beliefs, expecting certain magic or miracle to come from

Heaven in order for God to do for us what we ought to do for ourselves. The scripture, in the book Isaiah 65: 11 reads "But you men are those leaving Jehovah, those forgetting my holy mountain, those setting in order a table for the god of Good luck and those filling up mixed wine for the god of destiny". This is the proof that God acts within a process and a purpose and so expects mankind to do the same. Anything short of precision, order, variety, intelligence and beauty is short of nature's requirement. The holy writes themselves are a respected example of creativity. It was inspired. It is appreciated.

Furthermore, it is written in Philippians 4: 8; "Finally, brethren, whatsoever things re true, whatsoever things are honest, whatsoever things are just, whatsoever things are pure, whatsoever things are of good report; if there be any virtue, and if there be any praise, think on these things".

Do we substitute prayer for work? ***The book of Philipians 2: 12-13 reads..."Work out your salvation with fear and trembling". For it is God who worketh in you both to will and to do of his good pleasure",***

Daniel 1:4; children in whom there was no defect at all, but good in appearance and having insight into all wisdom and being acquainted with knowledge, and having discernment of what is known in whom also there was ability to stand in the palace of the king......"

About the talents and gifts in a person study the following scriptures;
Proverbs 18:16; "A man's gift maketh room for him, and bringeth him before great men.
Proverbs 10:4 "He becometh poor that dealeth with a slack hand, but the hand of the diligent maketh rich."
Proverbs 13:14 "The soul of the sluggard desireth, and hath nothing but the soul of the diligent shall be made fat".
Proverbs 21:5 The thoughts of the diligent tend only to plenteousness, but of everyone that is hasty only to want."
Proverbs 6:6 - "Go to the ant, thou sluggard; consider her ways and be wise".
Proverbs 22:29 - Seest thou a man diligent in his business? He shall stand before kings; he shall not stand before mean men".

Thessalonians 4:11 "And that ye study to be quiet, and to do your own business, and to work with your own hands as we commanded you".
1 Corinthians 7:7 " For I would that all men were even as myself. But every man hath his proper gift of God, one after this manner and another after that".
1 Corinthians 12:4 - "Now there are diversities of gifts, but the same spirit".
1 Corinthians 4:14 - "Neglect not the gift that is in thee".
James 1:17 - Every good gifts and every perfect gift of lights, with whom there is no variableness, neither shadow of turning"
Daniel 12:4 "Even to the time of the end many shall run to and fro, and knowledge shall be increased."
Proverbs 5:15: Drink waters out of thy own cistern, and running waters out of thy own well.
Verse 16: Let thy fountains be dispersed abroad, and rivers of water in the streets.

With the above scriptures, it is clearly evident that God, the Almighty creator understands the principles of creative potentials which He alone had endowed man within his own inside. The strength and purpose of talents in a person is to profit and make a source of blessing to others.

Let us consider these following strategic scriptures.

1 Corinthians 12:7. "But the manifestation of the spirit is given to every man to profit withal."
The clear essence of creative endowment is for profit and development. See the following: Matthew 25: 14-30
"For the Kingdom of heaven is as a man travelling into a far country who called his own servants, and delivered unto them his goods. And unto one he gave five talents to another two and to another one; to every according to his several ability; and straight way took his journey. Then he that had received the five talents went and traded with the same and made them other five talents. And likewise he that had received two, he also gained other two. But he that had received one went and digged in the earth, and hid his Lord's money.

After a long time the lord of those servants cometh, and reckoneth with them. And so he that had received five talents came and brought other five talents; behold I have gained beside them five talents more. His lord said to unto him, well done, thou good and faithful servant; thou hast been faithful over a few things, I will make thee ruler over many things; enter thou into the joy of thy lord.
He also that had received two talents came and said, Lord thou deliverest unto me two talents; behold I have gained two other talents beside them. His lord said unto him, well done, good and faithful servant; thou hast been faithful over a few things, I will make thee ruler over many things enter thou into the joy of the Lord.

Then he which had received the one talent came and said Lord, I knew thee that thou art an hard man, reaping where thou hast not sown, and gathering where thou hast not strawed: And I was afraid and went and hid thy talent in the earth; lo, there thou hast that is time. His lord answered and said unto him, thou wicked and slothful servant, thou knowest that I reap where I sow not and gather where I have not strawed. Thou oughtest to have put my money to the exchangers and then at any coming I should have received my own with usury. Take therefore the talent from him, and give it unto him which hath ten talents.
For unto everyone that hath shall be given and he shall have abundance but from him that hath not shall be taken away even that which he hath. And cast ye the unprofitable servant into outer darkness; there shall be weeping and gnashing of teeth."
It is clearly evident that certain responsibilities are expected of us to perform. God does not blame any other person for one's problems because He expects us to reach out to Him for help through digging deep within our own potentials.

The scriptures also recognizes specialized knowledge and witty inventions.
Hosea 4: 6(a); My people are destroyed for lack of knowledge. Apostle Peter was a fisherman, preacher and writer, while Apostle Paul was a lawyer and tent maker, then Jesus a teacher and carpenter.

Solomon, Daniel and a host of others who eventually became great in the scriptures were remarkable for their special skills and its applications. By this, they make their marks on the sands of time.
Then Job 32:8 caps it ; but there is a spirit in man and the inspiration of the Almighty giveth them understanding.
This recognizes the special spirit of God (characteristics and endowed abilities) ; inspiration - the power of realization; and understanding (the attribute of special perception intuitively, in order to achieve a purpose.

Job 34:4; let us choose to us judgement: let us know among ourselves what is good (this implies choice and desire for individual and collective good).

SUMMARY

1. Every thing we see and learn were created.
2. Man is a product of creativity.
3. Man names the animals and other things and has continued to name till this day.
4. Man created by a creator, has the mandate to create.
5. The creator of man is a great intelligence and has inbued man with this quality making man the most advanced being ever.
6. The strength of man to solve his problems and achieve utopia is only realizable by his continuous trace and reverential worship of his creator.
7. The strength of man lies in his mind which must be constantly purified by the creator; worship of any other thing contravenes his true essence.
8. The true spirit of creativity is that which develops its very strength from the spirit of confusion.
9. The true essence of humanity is creativity, promoting service through resourcefulness.
10. God creates and expects man to create positively.

CHAPTER 3

THE PSYCHOLOGICAL ESSENCE OF CREATIVITY

Aims/ Objectives:

1. To prove that psychology has its essence and perpetual search in creativity
2. To teach that creativity is an activity of the mind and the mind is the determinant of creative fulfillment.
3. To teach the renaissanced study of creative psychology.

The Psychological Essence Of Creativity.

The interest in people is more than just a religious practice involving preaching and ensuing peace here and there or more simply a desire to do good here and there, instead it is the far more outweighing towards correcting and a commitment to make this a life long affair.

This has informed the initial study called psychology. I have watched with keen interest, how people have viewed this field has a superstition or mere heresy.

For this, two reasons have been noticed.

1. The pioneering science of the mind is not viewed as realistic since it is viewed as an internal factor working with an internal mechanism, with less respect to its resultant effects on personality.

2. Because it challenges widely accepted materialistic views of man's personality and his existential place in the universe.

The psychologist is concerned with the science that deals with the mind, mental processes and behavior as well as the emotional and behavioral characteristics of an individual or group. A careful study of this field shows an unattractive nature because the strength of the mind thereby tending to lose the fact that man's existential place in the universe is his creative powers or endowments. The human mind is stronger than matter because it creates matter, it understands matter, it can alter matter and as well it has a complete control over matter. If the mind can be altered to develop positive attributes through proper orientation disseminated through information, then produced matter would only, predictably, be geared towards positive and productive ventures. The primary efforts of life should be directed, collectively, towards correcting the mistaken and narrow minded impressions regarding the issues of creativity in relation to the mind. I support the inner realization of the psychologist to understand the mind of man and I, hereby state, candidly that the real essence of man is in his reserved seat of universal creativity.

This is the beginning of the psychological analysis of issues such as telepathy, extra sensory perception, psychokinesis and what I shall summarize in my own terms as the psycho motive functions of the brain. Experimentally, examine the effects on the mind and then consequently on the brain of a person when simple positive words such as, " yes, we can "It is possible", "No doubts" "People, Rise and Shine", "never give up" in comparison to negative words such as " we're no better "our government lacks integrity" "demons are controlling our economy" there's no hope ", the effects of the initial are positive and confidence boosters, while words of the latter are negatives studded in a loaf and fed to destroy the mind and people who accept them.

I hereby state the laws of creative psychology, for anything studied learnt or researched without reference, in proper respect to the creative essence of such practice would do well to keep man in a state of perpetual purposelessness, fruitless search and internal bigotry of his true self.

1st law of creative psychology states:

"The strength and weakness of a person rests in the mind, with a resultant effect on the outward personality of such person".

This simply explains why people are prone to continuous learning, subject to environmental factors and different in bending towards attitude. The Greeks mentality of 'fight' has not changed till date, the British prayer of "God bless the queen" is perpetual in their minds, the Russians philosophy of "we're better" remains unquestionable, the Germans believe in themselves as the conquerors of Europe or moreso the German machines; the beauty of Paris in France, and the American "Yes, we can" of recent may be made in reference to Nigeria's "Good People, Great Nation" is more than just a Re-brand but an orientation, an eternal philosophy.

2nd Law of Creative Psychology
The essence of psychology is the harmony of the while person, while the essence of creative psychology is the promotion of human and universal peace, achievable by the appreciation of the purpose of existentiality.
History is the best keeper of records, and the most accountable of figures, since the end justifies the means. When money and funds are deliberately directed to destructive ends such as war, communal fights, nuclear aggrandizements, and fights; missiles and weaponry amassments, then history, is there to remind us of the worst of human activities during world Wars 1 & 2 and every other nefarious activities. The eventual tendency gears towards the stupid tendencies to show supremacy rather over others because you are afraid. I would rather advocate mental or intellectual show of supremacy than violence and wars. If everyone died, who would live. We need peace and must live in peace, peace, peace and more peace.

3rd: Law of Creative Psychology states:

> Man is not an animal, neither has he ever being one,
> but he is the creative representative of his creator,
> here on earth.

Analytically, man by the principles of darwinism has failed, but by the school of creationism is the only point of God's creation who was not ordered into existence but carefully made, totally different from the way animals or birds were. Man, this, has a purpose, a mind, a spirit, a nature but subject to the extent he understands the operation and the dynamics of his own internal workings.

KINDS OF CREATIVE PSYCHOLOGY

There are basically two kinds of creative psychology namely:

1. Systematic / predictive creative psychology
2. Parallel creative psychology

This involves the various methods which may be applied by the actively creative mind thought processes to determine future action. This applicative methods may be used by the individual, to foretell what results might occur for or against him in the course of his personal situations.
It would also involve the use of useful criticism, logical reasoning and common sense. Here the organizations would apply this compulsorily to aid effective planning. By the careful study of the past behaviors of a person or head, through various comparison of records it may be easier to deduce what kind of policy would be issued from the table of an organizational head or government office. A liberal minded individual would tend to introduce liberal policies, while a straight forward harry would prefer same when he is allowed an office of responsibility. Hence two major factors are considerable in carrying out this prediction method: (1) Personal attributes and attitude of a person and (2). Past records of such person.

The systematic or predictive study would help us in the following ways:

(i) To cope with stress and worries which may seem to come from disappointments divorce, infidelity, failures individuals and government policies as well as harsh economic realities.

(ii). To determine our own paths to avert, restrict or prevent certain disasters, by learning to study and overtake the situation far ahead ever before it occurs.

(iii). To take personal responsibilities for our own decisions and actions

(iv). To avoid so much talk and build the confidence to act or do;

(v). To appreciate our individual roles to each other in the society in which we find ourselves.

The strongest word in the process is "If" i.e. learning to assume certain consequences analyzed from present behaviours would not be referred to as half knowledge but preventive knowledge because invariably we are following the course of wisdom and learning to build a psychological shell around us; something like a shock absorber. History is full of people who refused to heed signs of imminent danger who payed dearly for it. The scriptures of Proverbs warns against the foolishness through simplicity without right doses of reasoning.

2. PARALLEL CREATIVE PSYCHOLOGY

This is the aspect of study which deals with the study of results based on previous actions, whether seen or unseen noticed or oversight. It deals strictly with detecting serious behavioural defects in a person. Such defects as previous crime as proved; working flaws such as laziness and lateness; poor record keeping of previous actions (as this is the characteristics of the modern age, for getting that our children would depend on them to make certain behavoural judgement especially if such incidents began to occur or make manifest in their own lives. Poor social and working relationships such as gossip; disregard and lack of respect to constituted authority and leadership. This study proves the inevitability of a person's past records running or trailing him in parallel lines to whatever office of responsibility he may ever hold, no matter how high that position way be. The greatest flaws or the best experience a person may ever have in his / her life is the quality of his home training. Proved; the blue print of a person's existential behavour would be found in his home training.

The plague of the modern civilization is the craze for ill gotten wealth, unaccountable money. In real situations, what do you make out of an organization which had existed in good financial stance, with good client or customer relationship, say, for the past forty years and then suddenly just three years ago, its reputation before the clients and customers began to fall, its fortunes dwindling and its over all standard falling?. It's not, I repeat, in the character of the assets or company name but in the character of its leaders and management. Leaders who have lacked integrity from the beginning perhaps unnoticed and have nursed the futile dream of heading this organization. With no mind of their own, poor understanding of the dynamics of mind operations, superstitions and self assuming, set of advisers who may also have the same kind of psychological problems he has, such an organization would quickly be rated and noticed because its fall would come quickly. The worst of these behavioural calumny would be the continuous shifts of blame from the top seat to those next until it gets to ground zero and then begins again to the top.

The leadership of a body corporate must be strong, resolute, understanding, practical and sincere. This is not limited or restricted to professional ethics but also must be boiled down to how, where and by whom did a person receive his home training, though with regards to environment. Not only for organization leadership also.

CREATIVE BONDING

The systematic or predictive and parallel creative psychologies are guided by a strong chain in every person called creative bonding. This would be better understood by the use of diagrams below.

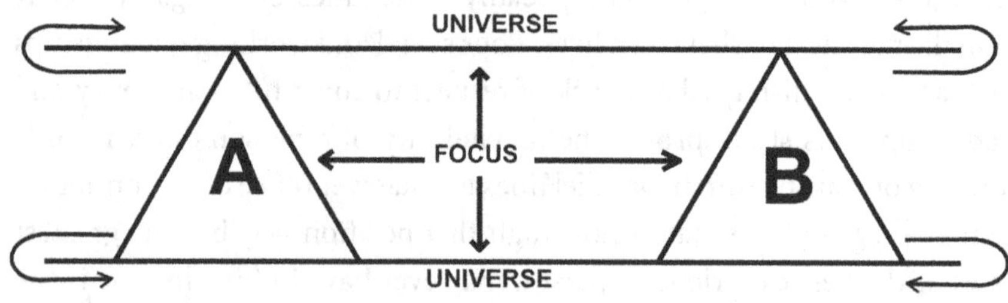

Diagram title: Point 1: Universal Relationships:

The point 1: universal relationships is a diagram of person - to- person (expressed as A+ B) bounded to their universal environment. The connective chain between the minds of person- to- person and then person to universe (environment) is the central point known as creative focus. This diagram proves the inter connectivity of the mind to the external factors in four cardinal points of upper existential faculties, lower existential faculties and then person to person mind factors in order to prove actions and determine results. Focus is thereby proved tò be the most electrifying and determinant factor in solution finding or personality predictive discovery. The greatests of creators, inventors, writers and philosophers have applied this ability to focus (seeing through their mind's eye to let the universe invisibilities come into view of their mental understanding. That is the creative power. The quality of the mind would depend on its informative ability to weigh and act accordingly. Moreso, the bond of creative focus is the point of individual concentration on an issue in order to get the right and desired results. This kind of focus is power fully electrifying as we shall see in the next diagram.

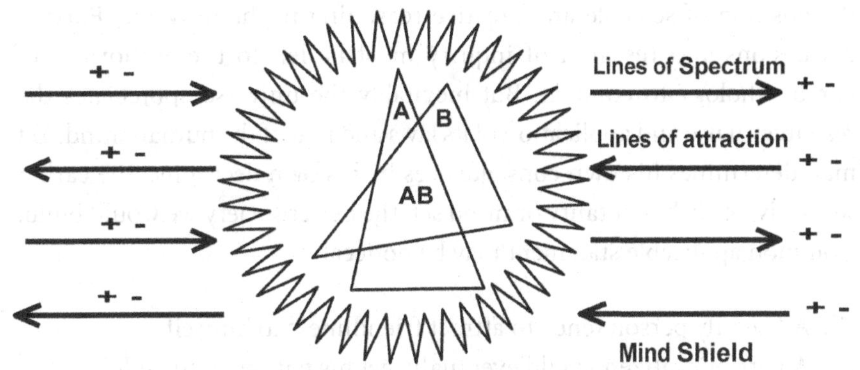

Diagram title: Point 2: ELECTRICAL AFFINITY.

Around the creative mind is a shield-like structure. This is called the mind shield. It is permeable and consists of the power to open or close as the possessor or the person may wish. This shield is opened as long as interest is developed in an issue and focus is maintained. This is the back up of the saying provable that the mind grows by what it feeds on, speaking in terms of the quality of information it allows in or out inexchange of its need to build either a fresh or to maintain a previous informative relationship. Focus is electrifying in effect. The lines of spectrum or focus are defined as electrical impulses of positive (+) or negative (-) nature proving the principle of "Like Attracts Like" in every form of thought processes. The lines of spectrum are magnetic in nature issuing from the focus of every person, while the lines of attraction are the resultant and opposite consequences and reaction issuing from the external factors or environment. These lines are magnetic in nature and must obey their own purpose of existentiality just as magnetic fluxes found in every sphere of the earth exerts power in their very sphere of influence e.g the force of gravity etc. From these findings, dreams will mature from thoughts, thoughts from concentration, concentration from desire, desire from choice into purpose. Better expressed in linear format as:

CONCENTRATION - THOUGHTS - DREAMS - PURPOSE - VISION - MISSION STRATEGY - TIMES ACTION - RESULTS

This overall understanding leads man to his point of either creativity or destructibility but the essence of creative psychology places man into the position of sensible and creative reasoning in the universe. Further discussions and research of inquiry must be left to the authorities of parapsychology to conduct. But inevitably the universe appreciates the resourceful use and application of its laws and that of the human mind. Yet man determines his own consequences by his own decisions. The earlier we cultivate such mentality or mind set, the better society we would build. Common applicable statement may be noticeable:

(1). A friendly person tends to attract friendliness to himself.
(2). A patriotic citizen would eventually get his nation on his side.
(3). What presentation a person makes before the mirror of the universe becomes the representations eventually returned to him.
(4). Minds who search for knowledge would eventually grow to become wise men.
(5). Success is never dependent on age but on the level of maturity to based on understanding.
(6). Those who try to destroy others would eventually destroy themselves.
(7). He who is self - indulgent, if not checked would become self destruct.

The laws of creative bounding also tend to join or bring people together into two categories:

1. Class of social bonding
2. Intuitive or Perceptible Bonding.

(1). CLASS OR SOCIAL BONDING:

This is defined as the group whose members have certain attributes in common; category. It is a division and bonding based on quality or grade. One is not fascinated when he finds or notices that the same mind set binds people of, perhaps, the rich class; or the elite class; or the average or even poor class. From these, greatness, I dare say, is a mentality. Poverty is also a mentality. No one would be surprised to find a picture of the creative geniuses, taken far back in the 1870's of four friends consisting Thomas Edison perched on an old mill Wheel, John Burroughs the famous naturalist; automaker Henry Ford; and Harvey Firestone whose rubber company was the largest in the world during an outing. (Reference Book; ALISTAIR COOKE'S AMERICA ; Page 255). Any one who would wish to join them would also be expected, logically, to work and live in their creative mentally. This is what I refer to as crossing-over; a situation where a person decides to join a class specifically by following or practicing an established norm or format from a previous class to another class.

(2). INTUITIVE / PERCEPTIBLE BONDING:

Intuitiveness would imply the faculty of knowing as if by instinct without conscious reasoning. A perception based on this impression and sharp insight. Such bonding would be a process where people of like minds come together by the influence of not social class but on certain understanding often not established. A sort of a mind affair. Perhaps understood when a rich prince desires a poor girl for his bride, or moreso, the passion of the rich for the plight of the poor; or mere attraction to a person describing such one special ans unique. Usually based on the individual judgement. Statements such as the following would prove this

(1). Beauty is in the eye of the beholder
(2). Opportunities abound in a seemingly less- developed economy.
Intuitive bonding seems more stronger, purposeful and useful. Noticeably, its characteristics have led to these particular findings which I have termed the unique strengths of intuitive bonding. These strengths are:
(1). Seemingly thinking pattern
(2). No distance limits

(1). SEEMINGLY THINKING PATTERN (STP):

The creative minds bonded by intuition seem to think alike. At some point in time, I have come to notice the ability to think what my friend or colleague or wife would think or be thinking almost at the time, thought at different definitive methods for the same or different reasons at different thinking frequencies but virtually for the same purpose or end. I have even thoughts to my surprise of perhaps what I perceive as my late father's thought and invariably found it in his books either expressed as simple notes by his own writing or specific to his own choice of authors. This is no magic but a special ability of creative minds because such minds seem to be open and sincere towards such issue It is however important that the creatively - alive mind is powerful when it is open (openness here would mean avoidance of filthy thought such as crime, dropping mind-blocking filths such as harboured guilts, faithlessness, doubts, hatred, unforgiveness and undue suspicion born out of the dreadest plague of humanity; fear). A mind which seeks to solve problems rather than search whom to blame for man's woes would tend to pick certain drops of light or specks of wisdom ideas. These ideas would even arrive in term of perceived signals: The mind of such persons at such times would not be independent. When ideas drop, as showers they tend to drop into the spirits of many persons at the same time for some, and at different times for others. The profit of such ideas would depend on what one does with it. Two weeks ago, in October, 2010. I received an unusual visitor. An educationist whom I had taught her daughter few years ago. During our discussion which she led I took particular notice of her thought patterns. I discovered that she felt pressured within her to pay me visits without even knowing the reasons. She is a Christian, so would simply interpret it as the Holy Spirit asking her to do so. But much to my surprise to realize that we thought alike as I shall summarize:

1. She expressed internal disaffection or dissatisfaction at how people had toward to religious piety and traditionalism substituting prayer for labour while their problems continued or persisted inadvertently.

2. Youths needed to be taught the right precepts and guided by the right examples to help the society produce sincere adults.

3. The Shocker: She had just purchased an exercise book four days earlier in order to conduct an experimental finding on the life- style of a turtle and named it; Creative sciences "the same title for this work which I have personally coined. Its beyond magic.

4. She was proving everyday to herself that demons were not enough distraction to stop us from reaching our goals if we simply trusted God thinking more about Him and then searching for the answers to our individual and collective problems.

I still believe that more research work perhaps, lay here for our experts of psychology. I am only concerned with the dynamics of the creative mind.

2. No Distance Limits (NDL)

I have observed that people with different thought pattern do not bond, and if they do, they would not last in such a relationship. Could this be the reason for marriage divorces business failures, crack down in friendships. Personally, I would study ought processes from samples of opinions and pick out of it; the pessimistic or optimistic ingredients, out of the individual's statements and then choose either to keep such and develop it; or stay totally out of it. I do not fancy pessimism or negativism because it hinders my creative thought processes.

Creative minds, provably, tend to think alike with no distance limits. They are strategists breaking every boundary of hindrances or obstacles until they eventually come across each other either in the family union, religious or secular relationships. Nevertheless this sort of mind set must be trained and developed to function for every one's collective benefit and happiness.

Talents And Prejudices.

The dynamics of the creative mind is the basic reason for studying the all important issues of the psychological aspects of creativity. We must take note of certain factors reasonable enough to guide us. The word "dynamics" for this study simply implies the social, intellectual, or moral forces that produce activity and change in a given sphere. This implies appreciating the inevitable power of development and reasonable change from a system of levels to another under the strategic guidance of productive thinking or reasoning. Change however is either moving in an upward or down ward trend or format. Yet change, the very fabric of the human mind is bombarded from all sides by certain in-ward abilities and choices. These inward abilities are called talents, potential, or endowments, while the choices I shall term prejudices.

According to the American Heritage Dictionary, the word talent means (1) "A natural or acquired ability; aptitude (2). Natural endowment or ability of a superior quality. (3). A person with such ability (4). Any of various ancient units of weight and money".
Also the literary oracle again defines the word prejudice as. "An adverse judgement or opinion formed before hand without knowledge of the facts.

This discuss is highly necessary because most people use them interchangeably resulting to poor personality development misuse of them.

Talents are natural to everyone. Talents are endowments which are subject to development and proper development. This feature is the very reason why they are called potential. A potential is something "capable of being, but not yet in existence; latent. It has the inbuilt capacity for growth, development, or coming into being. Compare it with its source word potent, which is something, body or system possessing strength powerful exerting or capable of exerting strong effects; Back in history the word talent was used to denote currency units and weights. When a talent is discovered and developed in a person it becomes a provable medium of exchange for life's dealings and profits. It becomes when traded a service meeting the need of people, creating opportunities as well as utility or satisfaction for others.

If for instance five hair stylists performed their vocation in a street the chances are greater proved by the following facts:

(1). Some will perform better than the others.
(2). Some will lead new strategies to lead the competition.
(3). Some would raise price for less quality while others would offer high price for good quality or reasonable price for good quality.
(4). Chances are that innovative ones would tend to perform better.
(5). Some would be referred to as gifted, talented or passionate (in terms of fulfilment) than the others.

Talents are not subject to flair or discretion.

They are undemocratic in operation because they have the ability to expose themselves without yours or anyone else's consent. Talents are the gifts given to you by your creator to profit, without which you would not find your place in life. The lack of respect for the endowment in herent in persons has led to the outrageous cases of unemployment, poor productiveness, inefficiency, mis-priority of careers, purposelessness and lack of fulfilment in life. One thing noticeable, is that the creative mind of man is an inexhaustible storage of talented deposits.

Prejudices have been interwoven wrongly into the issues of talents. While talents are discovered out of in-born interests, prejudices are set of likes and dislikes, sometimes, made out of ignorance or incomplete knowledge of the necessary facts. Interests are genuinely determined by only the individual, but prejudice is subject to environmental factors considerably determining the strength and weakness of every person.

The issues of talents have been mistakenly viewed from a narrow angle and myopic point of understanding. Talents have the spiritual thing in them often letting you have the right compass direction of your life. It Is independent of religion sex, race, or even training. You are unique because of the ability to express your own originality through your own talents. Following is a table of the observable qualities features or characteristics of talents and a set of prejudices.

Characteristics between a talent and a prejudices
(Summarizing the differences and similarities

	TALENTS	PREJUDICE
1.	It is God - made and influenced	Man made and environment influenced
2.	It is easier to develop and harness	it is more tasking and takes a longer time to develop.
3.	Cheaper to develop	Requires more resources.
4.	Promotes self dependance	Controlled by people's feelings.
5.	Its works sphere seems more a hobby	It often has difficulties meeting with with its purpose.
6.	Often occurs in original naturalness and uniqueness	A prejudiced job or flair fears its own expressiveness.
7.	It continues to be innovative	It is afraid of change.
8.	Workable without training formally	Requires stringent training methods
9.	It's non discriminatory	It's discriminatory especially on account on sex & race
10.	It brings fulfilment of its own merit, financial moral etc	Its tends towards egoism and eventually lacks true happiness and fulfilment.

It is important to develop our most fundamental resource which I believe is human resource i.e man power. Enormous opportunities are lost every pico second compared to whatever social aggrandisement or success we may think of ourselves. Well, what is the judgement:. Creativity has ruled the world and has made the greatest of its heroes for history's table of fame. However everyone is a hero if he can develop his talents.

CREATIVE CULTURE & PERSPECTIVES.

The human story cannot be complete without the discussion, or at least a special reference in psychology to their culture. Culture may be defined according to the American Heritage Dictionary as: "The behaviour patterns, arts, beliefs, institutions and all other products of human work and thought, especially as expressed in a particular community or period. 2. Intellectual and artistic activity and the works produced. 3. Development of the intellect through training or education."

Culture may also be simply defined as the social environment. We may further our study of this subject matter when we define creative culture. Creative culture is a product of creative psychology which distinctly studies culture as a relationship to creative applications for its own survival. This implies that creativity would thrive on the social background of acceptable beliefs, sensible religion, respect for ethics, and loyalty to proper education and understanding.

Creative culture is guided by two factors.

* Behaviour

* Values.

*Behaviour: Behaviour is the manner in which ones behaves; deportment. It is also the actions or reactions of persons or things under given circumstances. The behaviourism is a school of psychology that studies observable and quantifiable aspects of behaviour but not subjective phenomena. This would often prove the level of comportment a person has would readily show how he views issues and reacts to results. However while culture relates man greatly to his environment, creative culture proves its existence of the perspective of the influence of the social mind rather as the result of social environment and circumstances affecting man. The attitude of a person is often the results of their orientation. Orientation would ready imply the impact which a certain information would have on the understanding of this social mind. Creative culture seeks not to redress behaviour distinctively but to lead it to where it ought to be.

Behaviour can be induced. From this we may understand why the media must be used productively to promote the issues of proper behaviour based on the knowledge of the inner man. Behaviour is often a product of the instinct - the internal understanding that certain results would arise from certain actions and its counter reactions either in support or against such results. Take for instance a person involved in an orgy specially in sexual activity or any other activity.

On simple enquiry one of these would be noted.
(1). Perhaps he does not know its limits.
(2). Perhaps he finds inestimable pleasures from it.
(3). Perhaps it boosts his sense of egoism.
(4). Perhaps he views the opposite sex as meant for such a purpose.

From whatever line of thought we attempt to view it, we would discover a strong relationship between his environment and his orientation. The environment at factors would include the media - television, radio, music, movies and books, his home training and the company he keeps.
Orientation here would quickly imply his mind set. If his mind feels good about it, he would continue in such act until adverse effects force upon him a change of attitude, I shall then define behaviour as two types namely.
1. Animal Behaviourial Instinct (ABI)
2. Humane Behaviourial Instinct (HBI)

1. Animal Behaviourial Instinct (ABI): The ABI is easily provable in the circumstance of social extremes. This does not support the school of evolution which teaches that man was once existent and developed animals at various times of prehistoric developments. It is also wrong to think that man is an animal (whether higher or lower) simply because he has flesh. Going by the flesh principle, could it be sensible to learn and teach that plants are a special breed of animals simply beçause biology teaches us that they possess tissues, cells and xylem. Animals are guided by instincts which result to extremes while man is not subject to his instincts but a

thinking mind. The ABI is that pattern of behaviour which tends man towards senseless behaviour such as the instinct for war, carelessness, lack of respect for what is good, extreme religion fanaticism and general over indulgence. We are lowered to animals when we seek wanton power and dignity is thrown to the dustspan. These collectively may be referred to as animalistic tendencies because they disrespect true respect for lives, ideals and proper civilization, and they devalue the real essence of a credible and applicable value system. Such tendencies must be avoided through and judgmental, without a working formula for imbibing the ideals of a creative culture through integrating it into our social mind through the education system and professional life system.

2. HUMANE BEHAVIOURAL INSTINCT (HBI)

The word "Humane" refers to acts of compassion, being kind as well as emphasizing humanistic values and concerns. Such instinct would be the product of a positively - trained mind (PTM). It would emphasize the presence of a greater intelligence behind all intelligence, moderation in pleasure, modesty in fashion and attitude; it would encourage merit against bribery: correction against corruption: freedom for rightness, develop rather than depend, ideas rather than violence and everything good tending towards national development and human dignity against social ills. The HBI is one necessary to develop vast opportunities from every resource with regards to mind restructuring and social construction. It emphasis the results of decisions, the consequences of action and the power of choice. The HBI is a product of creative culture design.

2. VALUE

A value for this study, is a principle, standard, or quality considered worth while or desirable. It may also be seen as a worth measured in usefulness or importance or merit.

I shall define social values as a frame work which guide a people towards purposeful living and civilized thinking. This justifies the creativeness of the social mind of a people. It proves that the culture of a people is guided by what is referred to as the creative value system.(C.V.S). Since creativity screens social life individually and collectively, it proves that the CVS is measurable as either low or high, principally guided by the outward results. The people or nation with low C.V.S. would fundamentally be driven as hard of cattle, purposeless with over 65% unemployment rate, poor orientation, poor or no infrastructures, continual shifts of blame from citizens to government and government to citizens, poor sense of individuals and collective responsibility poor competitiveness, poor sense of patriotism social greed, social corruption, social ills of violence and extremes of religious practices neglecting the work ethics and national depression. The people or nation with a high C.V.S would live in awareness of these characteristics of the low CVS but would work to achieve a turn-around structure by introducing, developing, and maintaining a stable social mind structure individually and then collectively. Persons who study creative culture ethics are referred to as Culturists.

Culturists, guided by the creative ethics seek to redefine culture which may not be useful for our present age, while making recommendations for proper civilization and behaviour development to create better standards translated into the lives of the average person. To understand this rennaisanced field of culturist psychology, the distinction between culture designer and a culture destroyer is highly necessary owing to the fact that every system is either worked in the past, working in the present and would be re-developed, re-constructed or re-enforced in the future by someone. The major distinction between them is that while the culture destroyer stands against a culture, the designer works it.

DISTINCTION BETWEEN A CULTURE DESTROYER AND A CULTURE DESIGNER.

1. The culture destroyer sees himself different from the prevailing culture at a specified time, place for specified reasons while the culture designer would view himself as a part of the whole with a sense of responsibility towards its progress.

An inference to this would be viewed in a yard of people co-habiting with less than 10% of them highly educated but not financially buoyant to live in a better accommodation. Except there is a certain compromise, conflicts would also result from their level of understanding and perception resulting to fights.

2. The culture destroyer is an untamed designer while the designer is a tamed destroyer. Due to the untamed nature of the culture destroyer, he would have no sense of direction, although educated would have no support or friends and might perpetually die of his internal desires while being too critical, temperamental and judgmental, without a working formula for development or correction of the prevailing system. It is a fact that the poor and uneducated would readily gather their wits around a purposeful educated person whom they believe in and trust. This knowledge would readily serve itself hot on the menu of the culture designer who, though poor, has taken such advantage to correct the problems of these people (perhaps yard neighbours) who have a deep yearning for change and happiness. Yet, every analysis and solution formula must have the propensity to translate its worth to sensible values helping this society to eradicating hunger, unemployment, violence, depression and social ills. For practical applications, such person could be a haute couture (a fashion maker, perhaps a designer of exclusive fashion for women); a haute cuisiner (a chef elaborate and skillful at preparing food) and creating employment rather than being a hauteur (a haughty and arrogant person). This may sound fitting for local livers who are poor and malnourished. Such professions should at least give room for the participation of men and women openly, freely and professionally. Take for instance the stress relieving job of massaging, (the rubbing or kneading of parts of the body to aid circulation or relax the muscles) which encourages the man (known as the masseur) and the woman (called the masseuse) to work professionally and so many other vocations sensible and logical. The culturist thinking pattern applying it variously as well as in the intellectual circle would be noticeable when I bumped into the person of Masters Edgar Lee (1869 - 1950) who was an American poet. I have since wondered if, perhaps, he takes the credit for the creation of the Master's degree; an academic degree conferred upon those who complete at least one year of study beyond the bachelor's degree or else what I could do with a matinee; a dramatic or musical performance given in the afternoon.

The power of matter and material is not all laid in these physical phenomena but greatly in the mind of the user because whatever use, purpose or attraction he creates out of them has, and can only be defined by the mind of the user.

However, the present social ills of our age can be eliminated, unemployment can create more jobs than available labour; depression can become a reference impression and negative tides can be turned into positive outcomes if only we can begin to think and work in creative terms.

RELATIONSHIP BETWEEN BEHAVIOUR AND CULTURE.

1. SOCIAL NORMS OR CULTURE IS DEFINED LARGELY BY THE PREDOMINANT BEHAVIOUR OF INDIVIDUALS.

The culture of a society is so volatile because most of its partakers are subjected by certain environmental forces e.g. the media, to view certain people as the epitome of perfect living. These may be, in the view of the uneducated and poor learned, the leaders, the icons, the stars the achievers. We may not see the president's wife on trousers on a Sunday morning but sighting his daughter wearing such in the church might suggest to us that;

(1). The family has no fear of God;

(2). If the daughter of the president practices this without a complain from the pastor then it is morally acceptable;

(3). Though some people might murmur against this, but fear would not let them inform her father that the church does not welcome such norm.

The resultant effect of consistent attitudes of certain people in the society tends often to rub off on the prevailing culture of the people. For instance many people still nurse the fear of the democracy we preach and so do not even know where to tap or what to do to reap its benefits. This is because the predominance of harsh treatments by the former military governments still occupy their minds although they had never experienced anything personally with the military. This is what I refer to as the culture of fear. What messages are sent to the people by the government, media, elites and various organizations of great repute acts as the controlling factor of our culture control. The most controversial and influential psychologist of our time. B.F Skinner states. "No one knows the best way of raising children, paying workers, maintaining law and order, teaching or making people creative but it is possible to propose better ways than we now have and support them by predicting and eventually demonstrating more reinforcing results. This has been done in the past with the help of personal experience and folk wisdom, but a scientific analysis of human behaviour is obviously relevant. It helps in two ways: it defines what is to be done and suggests ways of doing it".

This statement confirms the need to develop certain simple, though complex and yet direct results of scientific analysis to defining human behaviour; through discovery of creative secrets and mysteries, to changing positively the present culture of poor existential definition, poor happiness through the acquiring of unsecured wealth, lack of trust for a government and leadership even if they are sincere and poor esteem as to the overall impact of citizens to their national life and humanity. The need for creative exploitation of the mind is the centre for which man's behaviour revolves and the culture depends. Get it right there and all goes well.

2. SOCIAL CONTINGENCIES MUST BE CHANGED TO AFFECT BEHAVIOUR AND THEN CULTURE POSITIVELY

The greatest duty of every society is to continue to reassure itself that all will be well. The problems of the society lies more at the neck of the elites. The problem is greater when we know that there are problems and become too comfortable to become concerned about deploying our most cherished resources to solving these problems. The understanding of this decision would be defined by the following questions: discussion

1. Can the leadership of a society, promise development without a tangible commitments to such ideal?
2. Does the present Ministry of labor have any useful bearing on the staff and factory workers of private organizations?
3. What happens if a separate Ministry of industry is created and incentives given to support the process of industrialization?
4. What about creating the Ministry of Creativity to oversee the creation and development of creative industries such as private museums, cartoon outfits, stage theaters, art and craft centers ?
5. What about encouraging the civilians to develop machines and master plans for professionalism of the law enforcement agents, the police and the military?

6. What about building nuclear power stations for electricity.

Understanding human behaviour will go a long way to developing the experimental analysis of our problems, its interpretations and its solutions.

It permits us to neglect irrelevant details, no matter how dramatic, and to emphasize features, which, without the help of the analysis, would be dismissed as trivial. This issue is not to be discarded as a mere reference ; just about talking old things in a new way; it's about setting the time target limit for breaking existing records. Contingencies occur in all aspects of our living and are most noted for the ingredients of measurable purpose and intention and they provide not just a mere abstract possibility but alternative formulations of so called social and individual "mental and practical processes."

The following principles would aid this process:

Principles of Contingencies:

1. Contingencies have interpretations, implying scientific studies and analysis.
2. Contingencies often lie unsolved, though accessible as a bridge between behaviour and culture.
3. Contingencies as problems must be anticipated and programs arranged to counter them.
4. Contingent behaviour can be changed by changing its controlling and functional conditions.
5. Contingent problems demand a reinforcement factor i.e creative strategy to keep alive the reviving mind its own keepers (culture).

Developing the New Creative Culture

According to a comedian friend of mine, Yawa Master he said " professors are different from great men because while the professors are made by the existing system, great men make new systems" but I dare say, agreeable that some great men, are worth professors and some professors are great men. This projects part of the conflicts existing in the present culture of humanity. These ate the testimonies held by B.F. Skinner in reference though unknowingly to the absence of creative utilization in developing the right culture or creative culture. The proper understanding and adapting this course study into our present and future educational system will invariably lead to the deeper appreciation of the human behaviour and thereby the ultimate creative age. This is the age where human talents are discovered through the analysis method, and developed through the influsion of the right message. Call it the freedom Age-one in which people can access their contingencies and use the available technology and means to change and control it for good; or call it the Info-time - a period when the massive information available on mind development would determine the success of a person within a short time and not age. This is what Ralph Waldo Emerson understood; what B.F. Skinner perceived and this what I have obtained.

According B.F. Skinner:

1. In trying to solve the terrifying problems that face us in the world today, we naturally turn to the things we do best..... As Darlington has said: Every new source from which man has increased his power on the earth has been used to diminish the prospects of his successors. All his progress has been made at expense of damage to his environment which he cannot repair and could not foresee.

2. The application of the physical and biological sciences alone will not solve our problems because the solutions lie in another field.

3. In short, we need to make vast changes in human behaviour...... (And there are other problems, such as the break down of our educational system and the disaffection and revolt of the young to which physical and biological technologies are so obviously irrelevant that they have never been applied.

4. What we need is a technology of behaviour. We could solve our problems quickly enough if we could adjust the growth of the world's population as precisely as we adjust the course of a space ship, or improve agriculture and industry with some of the confidence with which we accelerate high - energy particles, or move towards a peaceful world with something like the steady progress with which physics has approached absolute zero (even though

both remain presumably out of reached.) But a behavioural technology comparable in power and precision to physical and biological technology is lacking and those who do not find the very possibility ridiculous are more likely to be frightened by it than reassured. That is how far we are from "understanding human issues" in the sense in which physics and biology understand their fields and how far we from preventing the catastrophe toward which the world seems to be in exonerably moving.

5. Twenty five hundred years ago, it might have been said that man understands himself as well as any other part of his world. Today he is the thing he understands least. Physics and biology have come a long way, but there has been no comparable development of anything like a science of human behaviour.

6. It can always be argued that human behaviour is a particularly difficult field. It is, and we are especially likely to think so just because we are so inept in dealing with it. But modern physics and biology successfully treat subjects that are certainly no simpler than many aspects of human behaviour.

The difference is that the instruments and methods are not available in the fields of human behaviour is not an explanation, it is part of the puzzle. Was putting a man on the moon actually easier than improving education in our public schools? Or than constructing better kinds of living space for everyone? Or than making it possible for everyone to be gainfully employed and as a result to enjoy a higher standard of living? The choice was not a matter of priorities, for no one could have said that it was more important to get to the moon. The exciting thing about getting to the moon was its feasibility. Science and technology had reached the point at which with one great push, the thing could be done. There is no comparable excitement about the problems posed by human behaviour. We are not close to solutions.

7. It is easy to conclude that there must be something about human behaviour which makes a scientific analysis and hence an effective technology impossible but we have not by any means exhausted the possibilities. There is a sense in which it can be said that the methods of science have scarely yet

been applied to human behaviour. We have used the instruments of science, we have counted and measured and compared; but something essential to scientific practice is missing in almost all current discussions of human behaviour. It has to do with our treatment of the causes of behaviour. (The term "cause" is no longer common in sophisticated scientific writing but it will serve well enough here.)

8. Man's first experience with causes probably came from his own behaviour; things moved because he moved them. If other things moved, it was because someone else was moving them, and if the mover could not been seen, it was because he was invisible.

9. Intelligent People no longer believe that men are possessed by demons (although the exorcism of devils is occasionally practiced, and the diamonic has reappeared in the writings of psychotherapists) but human behaviour is still commonly attributed to in dwelling agents. A juvenile deliquent is said for example, to be suffering from a disturbed personality. There would be no point in saying it if the personality were not somehow distinct from the body which has got itself into trouble. The distinction is clear when one body is said to contain several personalities which control it in different ways at different times. Psycho analysts have identified three of these personalities - the ego, superego, and id-and interactions among them are said to be responsible for the behaviour of the man in whom they dwell.

10. Although physics soon stopped personifying things in this way, it continued for a long time, to speak as if they had wills, impulses, feelings, purposes, and other fragmentary attributes of an indwelling agent.

11. Careless references to purpose are still to be found in both physics and biology but good practice has no place for them, yet almost everyone attributes human behaviour to intentions, purposes, aims and goals. If it is still possible to show whether a machine can show purpose, the question implies, significantly that if it can it will more closely resemble a man.

12. Behaviour however is still attributed to human nature and there is an extensive "psychology of individuals differences in which people are compared and described in terms of traits of character, capacities, and abilities.

13. Almost everyone who is concerned with human affairs- as political scientist, philosopher, man of letters, economist, psychologist, linguist sociologist, theologian, anthropologist, educator, or psychotherapist - continues to talk about human behaviour in this present scientific way. Every issue of a daily paper every magazine, every professional journal, every book with any bearing whatsoever on human behaviour will supply examples. We are told that to control the number of people in the world we need to change attitudes toward children, overcome pride in size of family or in sexual potency, build some sense of responsibility toward offspring and to reduce the role played by a large family in allaying concern for old age. To work for peace we must deal with the will to power or the paranoid delusions of leaders, we must remember that wars begin in the minds of men, that there is something suicidal in man- a death instinct perhaps which leads to war and that man is aggressive by nature. To solve the problems of the poor we must inspire self respect encourage initiative and reduce frustration. To allay the disaffection of the young we must provide a sense of purpose and reduce feelings of alienation or hopelessness. Realizing that we have no effective means of doing any of this, we ourselves may experience a crisis of belief or a loss of confidence, which can be corrected only by returning to a faith in man's capacities. This is staple fare. Almost no one questions it. Yet there is nothing like it in modern physics or most of biology, and that fact may well explain why a science and a technology of behaviour have been so long delayed.

14. It is usually supposed that the behaviouristic" objection to ideas, feelings, traits of character, will, and so on concerns the stuff of which they are said to be made. Certain stubborn questions about the nature of mind have, of course been debated fore more than twenty-five hundred years and still go unanswered. How for example can the mind move the body? As late as 1965, Karl Popper could put the question this way: "what we want is to understand how such non physical things as purposes, deliberations, plans, decisions theories, tensions, and values can play a part in bringing

about physical changes in the physical world". And of course we also want to know where these non physical things come from...... A person's genetic endowment, a product of the evolution of the species, is said to explain part of the workings of his mind and his personal history the rest.

15. The function of the inner man is to provide an explanation which will not be explained in turn. Explanation stops with him. He is not a mediator between past history and current behaviour, he is a center from which behaviour emanates. He initiates, originates and creates and in doing so he remains, as he was for the Greeks, divine. We say that he is autonomous and so far as a science of behaviour is concerned, that means miracles.

16. The task of a scientific analysis is to explain how the behaviour of a person as a physical system is related to the conditions under which the human species evolved and the conditions under which the individual lives. Unless there is indeed some capricious or creative intervention, these events must be related and no intervention is infact needed... In fact, these dimensions of mind or character are said to be observable only through complex statistical procedures.

17. Before the nineteenth century, the environment was thought of simply as a passive setting in which many different kinds of organism were born, reproduced themselves and died. No one saw that the environment was responsible for the fact that there were many different kinds (and that fact, significantly enough was attributed to a creative mind). The trouble was that the environment acts in an inconspicuous way. It does not push or pull, it selects.

18. It is now clear that we must take into account what the environment does to an organism not only before but after it responds. Behaviour is shaped and maintained by its consequences. Once this fact is recognized, we can formulate the interaction between organism and environment in a much more comprehensive way. There are two important results.

One concerns the basic analysis. Behaviour which operates upon the environment to produce consequences (operant" behaviour) can be studied by arranging environments in which specific consequences are contingent upon it....... The second result is practical; the environment can be manipulated ... already well advanced and, it may prove to be commensurate with our problems.......... The result is a tremendous weight of traditional "knowledge", which must be corrected or displaced by a scientific analysis.

19. On the contrary many anthropologists, sociologists, and psychologists have used their expert knowledge to prove that man is free, purposeful, and responsible. Freud was a determinist - on faith, if not on the evidence - but many Freudians have no hesitation in re assuring their patients that they are free to choose among different courses of action and are in the long run the architects of their own destinies. This escape route is slowly closed as new evidences of the predictability of human behaviour are discovered. Personal exemption from a complete determinism is revoked as a scientific analysis progresses, particularly in accounting for the behavior of the individual.

20. By questioning the control exercised by Anthonomus man and demonstrating the control exercised by the environment, a science of behaviour also seems to question dignity or worth. A person is responsible for his behaviour, not only in the sense behaves badly but also in the sense that he is to be given credit and admired for his achievements.

21. By questioning the control exercised by autonomous man and demonstrating the control exercised by the environment, a science of behaviou also seems to question dignity or worth. A Person is responsible for his behaviour, not only in the sense that he may be justly blamed or punished, when he behaves badly, but also in the sense that he is to be given credit and admired for his achievements.

22. A scientific analysis shifts the credit as well as the blame to the environment, and traditional practices can then no longer be justified. These are sweeping changes and those who are committed to traditional theories and practices naturally resist them.

23. Freedom dignity and value are major issues and unfortunately they become more crucial as the power of a technology of behaviour become

more nearly commensurate with the problems to be solved. The very change which has brought some hope of a solution is responsible for growing opposition to the kind of solution proposed. This conflict is itself a problem in human behaviour and may be approached as such. A science of behaviour is by no means as far advanced as physics or biology but it has an advantage in that it may throw some might on its own difficulties.

Science is human behaviour and so is the opposition to science. What has happened in man's struggle for freedom and dignity and what problems arise when scientific knowledge begins to be relevant in that struggle ? Answers to these questions may help to clear the way for the technology we so badly need.

24. We often talk about things we cannot observe or measure with the precision demanded by a scientific analysis, and in doing so there is much to be gained from using terms and principles which have been worked out under more precise conditions.

25. When we have observed behavioural processes under controlled conditions, we can more easily spot them in the world at large. We can identity significant them in the world at large we can identity significant features of behaviour and of the environment and features of behaviour and of the environment and are therefore able to neglect insignificant ones, no matter how fascinating they may be.

26. The instances of behaviour cited in what follows are not offered as "proof" of the interpretation. The proof is to be found in the basic analysis. The principles used in interpreting the instances have a plausibility which would be lacking in principles drawn entirely from casual observation.

27. Amost all our major problems involve human behaviour and they cannot be solved by physical and biological technology alone. What needed is a technology of behaviour, but we have been slow to develop the science from which such a technology might be drawn... and with it possibly the only way to solve our problems".

I shall add therefore, that if the design for a better place of man's existence is to be contemplated let it be the design of a creative culture, the attitude to be whole.

SUMMARY

1. The science of psychology is not a teaching of superstitious beliefs, but involves a systematic science and study.
2. The science of psychology in relation to personality metabolism searches perpetually for man's place in the universal existence.
3. Creative psychology place's man in his existential purpose in discovery of his personality, his interests and his talents and endowments.
4. Creative psychology is governed by the laws of universal existence.
5. The 1st law of Creative Psychology emphasizes the strength of the mind, 2nd law teaches the relationship of internal harmony to world peace, and the 3rd law proves that man is not an animal nor has he ever been one to be termed so.
6. There are two kinds of creative psychology:
   A. Systematic or predictive.
   B. Parallel.
7. Provable factors of creative psychology lie in the point1 point2 diagrams.
8. Creative bonding is divided into:
   A. Class or social bonding
   B. intuitive or perceptible bonding
9. Talent and prejudices are ingredients of the creative person with different characteristics.

CHAPTER 4

CREATIVE PHYSIOLOGY & MEDICINE.

AIMS\ OBJECTIVES:

This study aims to achieve the following:

!. To expose the student to creative aspect of psychology and medicines.

!! To proves through creative thinking the connecting process between physiology and medicine.

CREATIVE PHYSIOLOGY & MEDICINE.

Physiology is defined as the biological science of the functions, activities, and processes of all the living organisms., and furthermore the functions of the organism.

Medicine is defined as the science of diagnosing, treating, or preventing disease or bodily injury: the branch of this science encompassing treatment by means other than surgery,an agent used to treat disease, often unpleasant but necessary or unavoidable

In understanding the scope of creativity to this aspects of study, it is necessary to take note of the following words and their definitions:

Physical - of or relating to the body, materials, or things with respect to the energy or the sciences dealing with them.

Physical Education - Education in the care and development of the human body, stressing athletics and including hygiene. Physical Examination - a medical examination to determine the condition of a person's health or physical fitness. Physical therapy: the treatment of disease and injury by mechanical means such as heat, exercise, light and massage, often done by a physical therapist.

The connecting thread between physiology and medicine is in the ability to understand the functions, activities, and precess of matter or body and then finding ways to prevent or treat certain malfunctions and the results of such malfunctions called or named disease. Interestingly psychiatry exists in medicine.

Creative physiology is defined as the branch of creativity which studies the functions, activities and processes of a body, matter or system with a view to perceive certain interval behavioural characteristics either fully developed an unutilized or undeveloped as a result of over sight. This being so, creative medicine is the study of materials perceptively with the view or intention to creating the right suggestions to making the right medicines.

The creative mind or scientist involved in this aspect of study fully may have been trained in this field. However, our health is a basic function of ourselves and it is necessary that we take notice of certain behaviors which we may experience from time to time.

The creative mind who has not been trained in this speciality may perform his creative physiological functions by observation, study, perception and note taking. He must begin by asking the right questions. This may be in regards to simple animal behaviour e.g his pet (a dog, cat), wriggling worm, a swimming turtle or fish a jumping frog or toad, the effects of dirts on an environment, trip hazards which may lead to domestic accidents. In simple creative medicine, he may wish to study perceptibly the the effect of balanced diet on children, applicable mixture of foods and their health effects based on the knowledge of their constituent vitamins and minerals; understudy of the constituents of some drugs when compared to another. For instance, why do drugs which contain chloroquine when taken alone, scratch the body while paracetamol helps me sleep and thereby reducing pain? Certain domestic questions such as:

(I) Why do I feel so much pain in my bowels (stomach) when I eat beans leading to purging: yet it pleases my taste.

(Ii) Why do I tend to urinate so much when even I have eaten lots of fruits or certain leaves in soup: for instance, I have the conviction that eating lots of bitter leaf soup helps my system to dissolve and release accumulate sugar or sugary substances from my body through the elimination process.

(Iii) Why does kerosene tend to stop the aching pains of a soldier - ant's sting?

(Iv) Is the issue of fatness or slimness a question of a person's nature or heredity?

(V) Why would the traditional bean cake (local moi moi or akara - beans) but soya beans would not cake that way yet they both contain protein? Could it suggest the presence of an amount of starch in beans, perhaps, rather than in soya beans; based on the experience that corn o r grounded maize does so when boiled in a nylon for a meal?

(Vi) Could there be some other health benefits, e.g. Curve to certain ailments from these plants we call grasses, perhaps could the curve to ailments such as cancer, hepatitis, HIV/AIDS, appendicitis and fibroid (without surgery) be discovered from simple plants, perhaps, presently neglected. What about insomnia and diabetes.
Could certain results of predictable certainty be made, perhaps to help activate and inspire the expert in physiology and medicine. Discoveries, here, made by the untrained physiologist or medic may be published as an observed system or deduction.

(Vii) What other effects does smoking have on the health of a person? Could there by chance be an effect of inhaled contents through smoking on the brain, perhaps on the memory or judgmental ability?

(Viii) Apart from the widely-accepted belief of Spiritual healing of the olive oil on a person, what real effect does it exert when swallowed by a person.

(Ix) What overall effects would be felt on issues such as aging and longevity when fish, fruits or other meals are eaten either half-cooked or completely cooked.

(X) What, for instance, would be the attitude of people who ate specific foods so much; for Instance:

Pepper spice - pessimism
Sugarless meals - pessimism

Sugary meals (sweetened meals) - optimism.
Water - optimism.

The creative mind is also interested in issues of detailed activity. If you discovered that a fellow person or student suffered paralysis, what would be your reaction? Mockery or sympathy, superstition or research? Fear or sensible confidence? I discovered that certain ailments resulted from the total stoppage or impairment of activity of a particular organ in the body of its sufferer. Such diseases include:

Paralysis - Loss or impairment of the ability to move a body part; loss of sensation over a region of the body; total stoppage or severe impairment of activity.
Allelopathy - The inhibition of growth in one plant by chemicals produced by another plant.
Fit-Medic - A seizure or convulsion especially one caused by epilepsy; the sudden appearance of a symptom such as coughing or sneezing: a sudden out burst etc.
Ischemia;- A decrease in the blood supply to a bodily organ or part caused by constriction or obstruction of the blood vessels.

Of course, there are many more. Yet, the interest of the creator mind (psysiologist, scientist, or medic) must be embedded in a knack for details. If something resulted as an abnormally or a disease from a simple diagnosis of an obstruction to a body activities, then what should be done (perhaps a form of exercise), what kind of food or substance should be eaten (dietary requirements), what sort of prescription should be suggested (drug administration) in order to prevent such an obstruction or careless disturbance of the activity or process already actively in place. Can there reasonably be a complete mixture of constituents into one drug called a panacea -a perceived remedy for all diseases, evils, or difficulties; a sort of a cure - all? Perhaps, hydrotherapy should be further studied. I read with excitement how Ben Carson was able to separate Siamese twins successful using his own devised methods, and Louis Pasteur (1822- 95), a French chemist invented the process of heating a beverage or other food, such as milk or beer, in order to kill micro-organism that could cause disease, spoilage, or under sired fermentation; this process, today is called Pasteurization named after him.

EFFECTS OF CREATIVITY ON HEALTH

Health is wealth. When there is health, there is wealth. But then what would we suppose when there is health and yet there is no wealth but poverty, sorrow and boredom. This is where creativity again is lacking since even a dirty environment is an indication to poor creative health applications and thinking. We must remember that germs are creatively working to cause man many problems to his health and man may make use of them as well to cure the diseases caused by them.

The effects of creativity are stated below.

1. Healthy Exercises: Creativity involves working which would include motion e.g driving, riding, walking, which is beneficial.
2. Feed Habit: A creative mind will seldom wish to eat negligently or carelessly because his mind is often preoccupied by useful findings and research results.
3. Over-indulgence: over-indulgence in food and drinks may impair the judgement of the creative person. Avoidance of these would keep him healthy.
4. Cleanliness: A creativity person's environment is expected to be clean, orderly and well ventilated.
5. Good Health: A creative person's satisfaction is in discovery. This fulfilment gives him happiness which helps him be come seldom ill, avoids loneliness because he is accompanied by his works; avoids boredom, because he always has something to think about. By this, he is generally happy, self-fulfilled and has a good living.

SUMMARY

1. Physiology involves the study of body functions, activities and processes of living organisms.
2. Medicine is implied by the various techniques employed by diagnosis to treat or prevent diseases.

3. The connecting thread between physiology and medicine is in the ability to understand the functions, activities and suggesting ways to treat diseases which could result to impair this normal activity.
4. Creative physiology involves the understanding of certain body functions and activities with a view to noticing areas of suggestive development resulting from our everyday activities.
5. Creative medicine involves the perception and application of creativity to developing the right medicines.
6. Trained physiologists and medics may apply the principles of creative thinking to their acquired skills while the general creative person may apply the functions of observation, study, perception and notes taking as suggestive areas of development to the specialist.
7. The effects of creativity on health include: healthy exercises through work - without stress; healthy feeding habits, avoidance of self indulgence; cleanliness and general good health.

CHAPTER 5

THE NATURAL ESSENCE OF CREATIVITY

AIMS/ OBJECTIVES

1. To help the student appreciate the innate search for a better existence in the heart of man.
2. To expound the various categories of study inherent (i.e science, art and commercial studies) as the various ways by which man has devised to understand creativity in nature.
3. To prove that the existence of a better life is possible.
4. To teach that former doubts of the possibility to lead a better civilization rested on the shoulders of ignorance and that the appreciation of creativity in the foundation of proper civilization.
5. To prove that change is dynamic, its process proves that the mind is the source of true happiness and peace.
6. To prove that education is not complete without the induction of the creative teaching.
7. To prove that creativity is natural ; and nature itself is the creation of a creative intelligence.

THE NATURAL ESSENCE OF CREATIVITY

The principle of the natural essence states that a people can actually attain a peaceful, prosperous and egalitarian society if they are willing to return to their natural essence who is creativity. This theory is supported by the prevailence of certain inclinations of the educated men, artists, scientists, economists and creationists. Their various beliefs spoken, were often considered too magnificient to be achieved. Let us look at them briefly.

(A). The Arts and Literary School.

Sir Thomas More (1478-1535) was known to be a man of bold imagination and vision. He also owe of the precursors of the renaissance - the new learning. More's Literary novel was called Utopia. In it, Utopia was described as an ideally perfect place especially in its social political and moral aspects. Though an irony of his own belief he doubted this holy inclination to achieve happiness since he saw it as an impractical idealist scheme. According to John Burgess Wilson, "We still use the word Utopia to describe the paradise that every politician promises the ideal world which men can build on reason charity and proper social organization ". Yet, Thomas More is credited to be a creative genius, and a pioneer of the field of historical writing; Despite the doubts and gullibility he could not identify the strength of creativity in achieving the state of Utopia since it could have applied to others as well.

(2). The Perfect Market Petrol.

The Economists once speculated the probability of a perfect market, a kind of a free economy. They thought about it, and believed it pessimistically. Then confessed it was not possible. They had not considered the theory of the national essence. By creativity people will produce freely and buy freely. Competition would enhance standard, quality and yet affordability.

(3). Creationism School of thought.

The holy Bible states:

(i). Matthew 5:48 "Be ye therefore perfect even as your Father which is in heaven is perfect.

(ii). Romans 12:2 "And be not conformed to this world; but be ye transformed by the renewing of your mind, that ye may prove what is good and acceptable and perfect will of God".

(Iii). Hebrew7:19 "For the law made nothing perfect but the bringing in of a better hope did; by the which we draw nigh unto God".

(iv). 1 John 4:18: There is no fear in love but perfect love casteth out fear; because fear hath torment. He that feareth is not made perfect in love.

(4). Science - The persistence of science over the centuries is merely hinged on the thought that the world can be rid of disease and unnecessary deaths by the application of certain and necessary natural laws. There is always room for competence because that is the essence of creativity and conformity to the proper use of potentials and endowments.

(5). Current Analysis/ Political Structure.

The current analysis of politics is the presuppositions by A.V. Dicey in 1864 of a system of government known as **Democracy.**

The American heritage dictionary renders it as:

1. Government by the people, exercised either directly or through elected representatives.
2. A political unit that has such a government.
3. Majority rule.
4. The principles of a social equality and respect for the individual within a community.

Democracy cannot be throughly practiced without the proper education, use, application and utilization of the individual's own potentials for the contributory development of a perfect people and governance. Today, democracy rules the world.

SUMMARY

1. A better life is achievable when the strengths of the economy is founded on the natural essence of creativity.
2. The principle of Utopia remains the dream of every literary mind,yet, it is supposedly non achievable. Its first developer was a man called Sir Thomas More (1478-1535), known as a creative genius and the father of historical writing.
3. The perfect market postulated by the Economist is a dream supposedly not achievable yet the present principles of marketing (door to door) proves its possibility.
4. The creationism school of thought prove that religion supports creativity.
5. Science means knowledge which justifies the existence for the search for perfect existence.
6. Politics through the years has tried the various aspects of governance and leadership which are creative thoughts of the human mind.
   Presently democracy was once postulated by A.V. Dicey; supporting its creative search for a better and peaceful society.
7. The various aspects of study support man's natural search for better living.

AIMS / OBJECTIVES.

1. To prove the large extent of creative consideration available.
2. To show that issues which relate to creativity are practical in every day life.
3 To prove that creative endowments are not myths but provable facts and physical evidences.
4. To teach the most modern applicable philosophies of creative abilities to careers, individualism and development of society.
5. To show the relationships in references to differences, uniqueness and similarities in creative matters.
6. To show by applicable philosophical thoughts, that every society is either wealthy or poor depending on the extent of emphasis placed on creative contributions of Individuals Citizens.

THE BASICS OF CREATIVITY.

This book may seem quite unconventional yet clearly, throughout the sacred pages of history, it has been proved that unconventional ideas may have solved or at least contributed greatly to solving the great problems of the world.

Henry Ford's mission of "democratize the automobile", Thomas Edison's fear of the dark led to his perfection of the electric bulb and today still receives credit for it, Graham Bell's invention of the telephone, the Wright brothers whose father as a bishop of the church contributed to the voices of those clamouring against the thought of flying, but never realizing that while he spoke of this as madness, his sons though bicycle repairers, were almost ready to test their first flying machine, funny enough the high -heel shows were invented by a woman so stout that she could not be kissed by her bridegroom on the wedding day; then Alexander Mills by his crazy imaginations invented the elevator because he needed an easier way to reach the top of the skyscrapers; Richard spikes invented the automatic gear shift, Joseph Gamelle invented the super charge system for the internal combustion engines, Garret A. Morgan invented the traffic signals. Then the madness -the same madness needed to bring out the best of creative talents and usher in civilization continued when Elbert E. Robinson made electronic trolley which was the precursor of the rapid transit system;

Charles Brooks needed to make information available to many people through readable pages so he invented the street newspaper invented the pencil sharpeners; the fountain pen was invented by Purvis John Love saw the need and William and the foremost pen by George A. Biro; so it's called a pen not biro; Lee Burridge invented the typewriting machine and A. Lovete he invented the advanced printing press.

The meals began to spoil and waste until John Standard invented the refrigerator; Sarah Boon invented the ironing board, George T. Samson invented the clothes dryer. At least we all bear witness to how much better life has become as a result of this "positive madness" as I shall refer to it.

All of these were possible because the seeds of "Developmental Strategy" were discovered. These seeds is what I personally refer to as "Creativity" or Creative abilities; somehow, somewhere, they had read something or had a direct experience with certain information. They were not self sufficient nor had their government designed the best economies for them. Most of these people though a list so endless, and few mentioned for the purpose of this book' were poor; so poor that they needed to do something fast; and timely too.

William Shakespeare - my personal role model, wrote " it is always fortunes normal manner to deny the poor man the benefit of a life time quest for wealth, It is from all these suffering by the poor man that I shall be saved," as well as my self too. Amen. I have said an Amen because I need to answer my prayer first before expecting God to do so. Bob Alonge added, "your desires in life will either encourage you or hunt you," Plato then spoke," things are only a couple of ideas, Ideas are more real than things. "But then creativity, that last hope of mankind's survival does come when you are angry enough, not to destroy lives nor its substance, but when it gets you going, like James Hardley Chase advised, "That is son,... get mad and stay mad. Your anger will keep you going when you run out of everything else. Just like mine is keeping me alive. Use your anger".

These lessons simply confirm the principles of enduring legacies surrounding the mystery of creating for lasting use and survival. But for the purpose of writing this book I shall not confine but discuss more broadly the issues of my personal research, discoveries which stretch its lines beyond my own very person, wrongly to be regarded as my personal opinions, and if considered so, then more of this onion should be split for personal understanding and direct application to the individual lives of our people, ourselves and our generations yet unborn.

Sincerely, I shall claim no bogous titles in this field of which I cannot defend the bogousness of such titles. I am no professor of literature. I am no scientist, but at least, I do have a sense of perception and discernment to discover my creative ability and work towards making this applicable, producing positive results and the necessary impact for good. I shall hereby, befit to myself, the humble title of a private researcher on ethics of living and thereby not on the wide scope of English literature, but on literary and writing skills. The sacred details of English nor creative writing are not known to me nor an award of a degree given in my name, but I sure, do know my tenses, my characters of imagery, my plot, my message, the theme and at least, how to dissect my opinion on issues through disseminating information by the use of the appropriate words

The issues of inspiration, imagination, research and other few principles shall be discussed herein, as I have experienced and studied them personally: I believe I wrote my first manuscript at seventeen., Compiled my first three thousand principle by twenty and developed my need to build a private sturdy by twenty- two. Married at twenty- four, had my first child at twenty-five and had discovered my purpose in life by twenty-eight. For the writhing field, I think that one who has spent at least ten years in play writing and other literary works, It would be of immense interest to speech my personal guide train and various experiences on paper, especially in simple terms. My life is a reality of creativity through writing. Writing is what I enjoy doing; talking aimlessly, bores me. The life of the writer is in his books. Shakespeare wrote on the issues of poverty, power and fame because he knew them firstly, by perception and secondly, by experience. The strength of the writer is in his pen and his weapon is the paper that slim sheet made from wood.

The basics of my entire research on writing which have kept me on-shore this strenuous terrain for writing is time consuming and stressful is hinged on the following words.

1. **ORACLE:**

This word used as the title of this book indicating by diagram "the light of knowledge and creativity" is one of the most commonly used in our traditional plays. You would often hear a character portrayed as a priest speak the words of the oracle; either for good or for doom to a people at a certain time. This word, one of my greatest appreciated words has a meaning in respect to search, research, enquire, find out about, find out why, what shall be done; how and in what manner should it be done; and this confirms the longing in a man's heart to lead a better life by knowing and applying. To know is knowledge; and to apply is wisdom especially applying profitably.

According to the Oxford Advanced learner's Dictionary, 6th Edition; Oracle is defined as

1. A place where (in ancient Greece) people could go to ask the gods for advice or information about the future, the priest or priestess through whom the gods were thought to give their message e.g. They consulted the oracle at Delphi.
2. (In ancient Greece) the advice or information that the gods gave, which often bad a hidden meaning;
3. A person or book that gives valuable advice on information e.g. My sister's the oracle on investment matters. Take note of the "person" or "book" in the third definition; as well as the words "hidden meaning" in the second definition; then the phrase, "to ask for advice" in the first definition.

The Collins Gem English Dictionary defines it as:

1. Shrine of an ancient god; prophecy; often obscure, revealed at a shrine; person believed to make infallible predictions.

Then the American Heritage Dictionary third Edition bares its mind:

1a. A shrine consecrated to a prophetic deity;
b. A priest or priestess at such as shrine;
c. A prophecy made known at such a shrine.
2. A wise person.

2. MUSE

A neglected word of significant impact not only for the writer but a creator such as an artist, musician, composer, perfector and a developer.

The Oxford Advanced learner's Dictionary, 6th Edition defines it as:

1. Person or spirit that gives a writer, painter, musician, etc, ideas and the desire to create things e.g. He felt that his muse had deserted him (that he could no longer write, paint, etc);
2. (In ancient Greek and Roman stories) one of the nine goddesses who encouraged poetry, music and other branches of art and literature.
   Collins Geml English Dictionary confirms:Muse as;
1. (Greek myth) one of nine goddesses, each of whom inspired an art or science;
2. Force that inspires a creative artist; while the american Heritage Dictionary says;
1. (GK. Myth) - Any of the nine daughters of zeus, each of whom presided over a different art or science.
2. A source of inspiration, especially of a poet. (GK. Mousa).
   Then, why wonder at the source of the word given as a name for a place devoted to the acquisition, study, and exhibition of objects of scientific, historical, or artistic value - that is the word; MUSEUM- from the Greek word Mouseion, shrine of the muses.

3. LITERATURE:

Collins Gem English Dictionary defines this word as:

1. Written works such as novels, plays, and poverty; books and writings of a country, period or subject. The American Heritage Dictionary defines it as;
1. Imaginative or creative writing.
2. The body of written works of a particular language, period, or culture.
3. Printed material of any kind as for a political campaign.

4. RENAISSANCE:

The American Heritage Dictionary defines it as:

1. A rebirth or revival.

2a. The humanistic revival of classical art, architecture, literature, and learning in Europe.
b. The period of this revival, roughly the 14th through the 16th century.
3. A revival of intellectual or artistic achievement. It agrees with the Collins Gem English Dictionary's definitive breakdown as;
1. Revival or rebirth ; revival of learning in the 14th-16th centuries.

5. THEATRE:

1. Place where plays etc are perform.
2. Drama and acting in general.
   (Collins Gem English Dictionary): then being defined by the American Heritage Dictionary as:
1. A building for the presentation of plays, films, or other dramatic performances:
2. A room with tiers of seats used for lectures or demonstrations;
3. Dramatic literature or performance.

6. CLASSIC

The American Heritage Dictionary defines this word of my liking as:

1. The languages and literature of ancient Greece and Rome; a typical example etc. whereby Collins Gem Dictionary says:
1. Being a typical example of something;
2. Of lasting interest because of excellence;
3. Attractive because of simplicity of form;
4. Author, artist or work of art of recognized excellence.
5. Study of ancient Greek and Roman literature and culture.

Research Result

On the careful study of these words, few issues, yet very pertained in the world of the creator and author of literature genres, such as myself; may be noticed and must be critically looked into for us to make practical sense of these issue.

The following factors are noticeably correct, with no restricts or coufecement to knowledge either of the Arts or science or perhaps, the

INTEREST:

The Greeks and the Romans had special notice of trivial issue. This built them an **INTEREST.** Such interest would later influence their general belief, research and behaviourial patterns towards issues bothering at a time- being. The creator and writer notice quickly; he is adepts at issue, matter which others withers concern would consider as commonitrivial, or minor, for instance, what influences weather? Why am I black or fair complexioned? Why do some trees grow so tall and gigantic? What about fire? Who made the earth? These people hasted the word, mystery; something which was hidden, profound unexplainable and no wonder they would spend time and much of their other resources to ask, find seek, enquire and most times, remained restless, hungry and patient until they found to the fullest the curses; or more the cause and effect and effect of the so called mystery. In creativity interest is first, others then follow.

The American Heritage Dictionary there fore defines the WORD "Interest as," A state of curiosity or concern about or attention to something' something that elopes this mental state" This implies that issues which arouse interest are often independent of academic qualification or social status; from these interest, mistakes are found and creations are made. I shall simply define interest as:**That point of catching the see mightily absent";** it's been there but no other saw it, and if they didn't then I was the first to become absorbed in it enough to make it better or modify it. They were interested. That made the difference. That simplest turned the globe from primitive age it used to be into the present civilization. Here, I must state my convictions concerning the unrealistic struggle for a better placing. Over the years by the adherents of the field of science and the Arts. In as much as one must make a Sider, it is necessary that both realizes their place, basically as creator structurally, socially morally and economically important to the society of people While the scientist tries to create the rainguage for the wet day, the artist draws a scenery out of it to dress the sitting room and the poet writes a poetry out it helping it become expressive for the understanding of even the less - lettered.

If science means knowledge technology for life's benefit, than arts would simply be, the expression of such knowledge, most honesty, some of the best legal cases have been won deducing them mathematically. In this book, as we go on, I shall prove such factors, inherently known; into mathematical formulae, for clearer memory and emphasis.

This shall be known as P.M.A.ku's law of.... P.M.Aku being the shortened format of the author's home.

Basically, interest is the result of curiosity within a specified time and place, with respect to incident which occurred at such time and place.

Pmakunatically: P.M.Aku's law of interest shall be

$$\text{Interest} = \frac{Cu \times Tm \times P_1}{Occ.} \text{ Or,}$$

$$\frac{Cu \times Tm\ P_1}{Occ.} = In, \text{ so}$$

$$\frac{CTP}{Oc} = I; \text{ that is the formulae for interest.}$$

But curiosity exists in levels, though not seen, it can be felt by the up surge or down surge of energy which produces Heat. So, I shall assign the unit C. joules: that the amount of energy measured in joules determining the extent of curiosity; since time and place are constant and can be variable also, we shall therefore assign 'Interest', a person has in something with the unit Cj.... Interest grows with thoughts. I shall also state that interest, borne of curiosity exists in what I shall refer to as the:

Thermometric Thinking Levels (TTL or T2L)

Which I break into three parts:

1. Low TTL
2. Medium TTL
3. High TTL

1. Low TTL:

This is the first time a question into an issue arises. Most times, it comes with the feelings of insecurity, defeatism; the person feels crushed; perhaps, for not knowing. For instance, in certain parts of Africa, there is a saying which goes thus "What one does not know cannot kill him," but, right now, I disprove such statement because it makes mockery of education. I would rather say,"What one does not know is greater than him". Ignorance is a folly.

The Low TTL often begins with the question "Why", "How", "I don't understand", and at this level of enquiry, the person with this level who shuns the audacity to enquire, quickly loose s interest and remains ignorant on such issue.

2. Medium TTL:

This person begins further search which more often than not demand his resources. He begins to find answers little by little, thereby arousing himself to go higher. You hear him says 'Wow'! 'I never knew that', 'I'm not sure', 'Let me try this way or that', his percentage of TTL rises.

2. High TTL:

This is the highest point of interest. Nothing can stop him. The secret of research are normal to to him at this level. He is at the stage where an idea can be developed.

Categories of TTL on percentage.

High TTL= 75% - 100 %
Medium TTL = 45% - 75%
Low (Grade ii) - 15% - 45%
Low (Grade I) - 0% -15%

The low grade is divided into two:
(i). Low Grade i
(b). Low Grade ii
(A). Low Grade

(ii). Must be considered first because the person possessing this level of interests is at what I shall refer to as the wishing stage. The wishing stage is the point of experience lack of knowledge on an issues while believing that someone else cannot help him, but if he could he could learn.

(B). Low Grade (i): I shall refer to this as the sleeping or dormancy level stage. Not death because, the possibility of aroused is still there a person at this level is outrightly unconcerned, totally regretful into eternally but outrightly stubborn outwardly. State wants to indicate such a person are the like of:

*. I don't want to hear
*. You cannot force me to listen
*. Keep your knowledge to yourself;
*. What is my business?

These set of person's tend to love creativity, but hate its process they are stuck to their own pattern of life and would resist change even if it were in their best interest. Most times, they experiences morbid fear out of sheer stupidity. Interest is necessary toil for the spring genius.

2. KNOWLEDGE APPRECIATION

Interest makes you search as you search you get and gather information on such topical issues. The Greek's and the Romans appreciation of this process led them to many realizations. The storage of this information is called **knowledge**. The appreciation of knowledge superceded undue criticisms of their economy because in poor economies, does Knowledge provide opportunities. One may question why some rich countries still suffer despite the vast knowledge laid at their disposal. This, I shall attempt to answer on what I call types of knowledge.

TYPES OF CREATIVE KNOWLEDGE

There are basically two types of knowledge:

(I). Present knowledge.
(Ii). Absent knowledge

Present Knowledge: This is the knowledge one has about an issue. This is the knowledge to be common e.g. The price of fish in the market may help

the house wife plan; but the cost of a particles or brand of car does not go street labels. There present knowledge may be divided into two:
(A). Present usable knowledge (PUK)
(B). Present non-usable knowledge (PNUK) -Pronounce as KN- O-K.

(B). Absent Knowledge: this knowledge is there, but not known, nor common and would require a further search or research in order to find it.
I shall divide this into two parts;
(A). Absent Intuitive Knowledge (AIK).
(B). Absent uncommon knowledge (AUK).

(A). Absent Intuitive Knowledge (AIK).
Follows the principle that your mind tells you about that thing. Commonly, if often seems an idea so unrealistic because it takes a highly trained mind to perceive strange ideas. Here the muse the internal power has a voice. I wish to state that I believe the muse here could be something like the Conscience or the super or powers one has submitted himself to. This kind of knowledge makes the mind provides sudden solution to problems which have all along been thought difficult.

(B). Absent Uncommon Knowledge (AUK).
The AUK is that knowledge which is lacking on processing and obtainance. It takes an uncommon bravery to go in search of it; for instances, certain operational secrets or business secrets lie only with uncommon to every person on the streets. Here, you would not be surprised that the American President Theodore Roosevelt was so caught in a book that he never realized it was raining; or more so Aristotle (384-322BC) Greek Philosopher sat in a swimming pool when he ran out naked screaming "Eureka" "Eureka"! Which means " I've found it! "I 've found it. "Truly, I confirmed also Pythagoras 6th cent.' Greek mathematician and philosopher discovered foreign and developed the Pythagoras theorem student today. Board a ship while I worked with then on board ships, as an Engine Room Assistant, here in Lagos Nigeria. Today that knowledge and discovery of the Pythagoras Theorem - made Aristotle - a great and famous man or do you also interest yourself to know that a great moment in science came when Albert Einstein discovered that time is actually money' or how his formula showed the

American government the extent of damage a bomb was doing to mankind when measured in mass and capacity. There you are.

P.M.AKU'S LAW OF KNOWLEDGE.

"Knowledge occurs within a certain experiences at a specified time and place." Time and place occurring together as a constant. Obtaining knowledge is dependent on one's decision and ability. Knowledge occurs in volume. Its unit shall be like litters present. Pmakumatically, I shall experience it as;

P.M.aku's law of knowledge
Kn x Ex
De + Ab = Tp or Ktp = Kn x Ex
De$ Ab

Where K represents constancy of time (T) peak of aspiration
And place cp;
De = Decision
Ab = ability

P.M.AKU'S TRIANGLE OF KNOWLEDGE

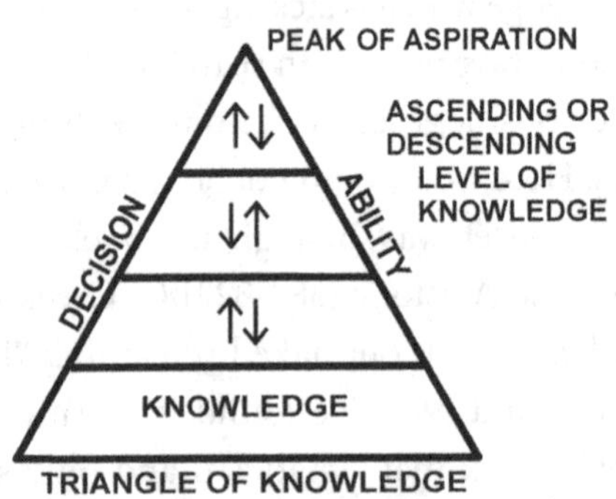

Knowledge grows when cultivated. It can also diminish by poor zeal or forgetfulness. Without knowledge, there can be no creativity, in matters of creative knowledge (ck); intelligence is better required than brilliance. The book worm may still lie in the castle of diamonds and yet see nothing but mind. Creative we citizens dictate to the government but clearly, they do not depend on the government. They are a people with a mind, not a herd of led cows; they challenge and obtain where they are; not grumble and complain.

There is need to search for more knowledge. The fruits tree with a medicinal effect; in school, home, everywhere. Remember," my people die for lack of knowledge-" the holy bible. A new book. I try to tell the story to a few people and then I note their reaction. through this method, I am able to listen to their various criticism, note how they perceive this story. I also conduct what I shall call SAMPLING. Sampling is the process of showing a few made copies of such product to my potential buyer or one who is involved in such a field. During oral research; respect and open - mindedness are two indispensable attitudes one cannot do without. Then it also pertinent to involve all ages and class; whether of the matured age - range: middle age -range and going ones if they seen excited or feel uninterested about your new ideal, then the first thing would before you to be guided by their responses into the next level of research.

2. INTELLECTUAL RESEARCH:
Here, books and previous samples of such work (s) is consulted. Presently the use of the internet is highly beneficial. Patience is the necessary tool. An Gagle-Eye must also be kept on every available resource so as not to let the necessary facts pass you unnoticed. The essence of having a keen watch is an order to keep focus alive. As much as possible, quietness and discretion are indispensable attitudes during point.
Sometimes, here, there are what I shall refer to as CATCH - PHRASES; which are statements of special interest to the work at hand or in the future. Here notes are taken. Analysis of problems are done with regards to its cause and effect borne in mind. The cause and effect 'principle helps you ask the necessary questions as to what caused or it lingered for so long; what impacts or effect has it had on the people and what are the factors to be considered if this problems must be tackled.

I shall broadly divide cause and Effect into two.
(a). Direct Cause and Effect.
(b). Indirect Cause and Effect.

(A). Direct Cause and Effect:
This is a method of inquiry into those factor which were directly responsible for such problem or occurrence. Such factors includes the following divisions;

(1). Physical Factor e.g weather elements such as rain, soil time etc.
(2). Economic Factor e.g government policies business trends inappropriate laws
(3). Social factors e.g people attitudes which are either positive or embracing; or pessinitive or result
(4). Religions Factors e.g. Praying or praying and working; or working instead of praying; or praying and working. The issues of religions activity by a people also goes a long way to effect their attitudes and opinions concerning an issues.

Personal Factors:
The me, Myself and I principle. An I egocentric or open-minded, by my action do I cause others to becomes inspired or expired; hopeful or despair, respectful or disdain. My opinion about society. Should show whether I am optimistic or pessimistic about my society.

B. INDIRECT CAUSE AND EFFECT;

This deals with those causes and effects which occurs as a result of actions considered trivial or not enough to cause problems but are yet to display their effects on the long term. For instance issues of global warming; neighborhood smoking; next door music disturbance, open refuses dump, drunk driving and public issues which effect each person and other collectively. If the world suddenly blows up as a result of collective emissions from everyone's chimney, then all is effected, yet those emissions or smoke seen too little. Such analysis are important.

3. PERSONAL RESEARCH.
This is the stage at which you have arrived at certain points of your research. This stages involves
* Thinking
* Deductive

* Conclusive
* Action

* Thinking Stage helps your mind see the major factors of the problem. A problem here would mean a scarcity or inadequacy, or perhaps certain over dependence on one aspect over others thereby causing a neglect in the system. Thinking must be sincere if the best results should come.

(B). Deductive Stage:

Here, the researchers begins to note those special and must important issues. These issues I calls specials because they are of urgent importance and must be looked into very quickly, if treated it is hoped that their positive effects would be over whelming.

(C). Conclusive Stages:

Here, conclusions are made personal reports are written and developed. Solutions are drench.

(D). Action Stages:

Solutions are fully developed with humanity in mind. Anything short of this is futile. In drawing an action plan, it is important to find a place where your own impact would be felt. What you would do if given an opportunity to change the system. Costs implication on individuals is important deperating on the class of people you are creating for. These costs must be reasonable, realistic and affordable.

4. USE OF THE MIND:

More beyond mere thinking or use of the intellectual faculty or calculations, the Greeks and Romans recognized. The power of spiritual influence on the mind. Somehow, man must draw nourishment and spiritual help from a more superior source.

I shall describe the mind as a central point between man and his creator. The bottom part of the mind is what I personally refer to as Treston. Psychologists do refer to a deep part of the mind as the sub conscious. I have come to realize that in the mind of a creator, there seems to be a point of reserves of creative resources but that which will be preferred solution to, would I have named requested. This reservoir is what I have named the word Treston. So I clearly define the word Treston the reservoir of creative resource in the mind of the creator or creative person. Personally I have experienced certain difficult situations. As a Christian I pray about it, believing deep down that somehow answers and possible solutions would come. Perhaps I could see it in the dream; perhaps it would occur in the bathroom. What you dwell your mind upon as I have proved has a rebounding effect, whether positive or negative. But the Creative person is an optimistic person, confident, hopeful, resourceful, hard and smart worker and persevering. It shows you the direction and points of possibilities even when it seems not possible. The Creator suffers a disease, I shall called Restless Creativity. I wish to suffer more of this disease.

The mind is the sub conscious part of every person. By natural philosophy, it is already endowered with certain potentials which are often unique to the person. The P.M. Aku's structure of the Creative mind is controlled by a Central Box of Understanding. This is the engine room of all thoughts and idea processes in the subconscious. Flanked on the left and right sides are the Central Entry Canal and the Central Exit Canal. These form the passage way of information obtained by the person to and from the external sources for processing in the B of U. Above the B of U is the Seeds of Ideas. These often varied remain above the person's grasp (often seem as a mystery) until it is through the vent of transmission to the B of U after which dependent upon other factors such as Knowledge, experience, desire and choice; and decision, a deep search for sincere solutions to the problem is sought from the Treston, either positively (+) or negatively (-). It is clear that here in the Treston, answers to questions lie here as Creative endowments. There is a constant travel to and from the treston to the B of U. The four powers of the brain; but here, they play a greater role of accepting and transferring their factors from the external point. The absorptive chamber attaches itself to knowledge, the Retentive attaches to external experience; the imaginative to Desire and choice, then the judgmental chamber to every decision working or wade.

The seeds of ideas are controlled by the thought pattern. These thoughts are, to a large extent, controlled by the environment. So the four chambers would naturally act as Control monitors of every thing or action going around the person. The double and opposite arrows occuring between the chambers and the B of U proves a reciprocal effect. Each acts in a contributory manner to each other, helping the individual become a more balanced person.

The Treston deep is the custodian of the ever flowing reservoir of endournments. This reservoir is endless and lasts throughout one's existence. Here this understanding reflects on both Physical and spiritual phonomena. Flowing is the ash of the deep. Interestingly, most genuiness of our world are those believed to have studied, tapped and developed superior creative powers outwardly when they were able to dig from the Treston - deep. The average person is most times, able to reach the treston fairly.

In order to understand these concepts better, let us study the following statements adapted from Harold Sherman's "How to picture what your want"

1. Owe to the most remarkable examples of the working of this Creative power of mind was demonstrated by Albert Einstein. Working as a poor patient office clerk in Berne, Switzerland, he had devoted every spare moment, night and day, for months in an attempt to develop a mathematical formula by which he could prove the inter-relationship of time, space and matter. Goaded by an unsympathetic wife, who urged him to give up his "fruitless quest" and spend all his time toward enquiring a better living, Einstein retired late one night, filled with a desperate now-or-never yearning.

Suddenly, in a deep dream state, a pahoramic image of the universe illuminate his consciousness! He awakened with the knowledge he had been seeking". The energy contained in any particle of matter is equal to the mass of that matter multiplied by the speed of light one hundred and eighty-six thoudsand miles a second - and again multiplied by the speed of light".

All these, Eintstein expressed in his now famous formula: "Equals MC Squared! Once possessed of this knowledge it has taken Einstein three weeks to put it down on paper !.

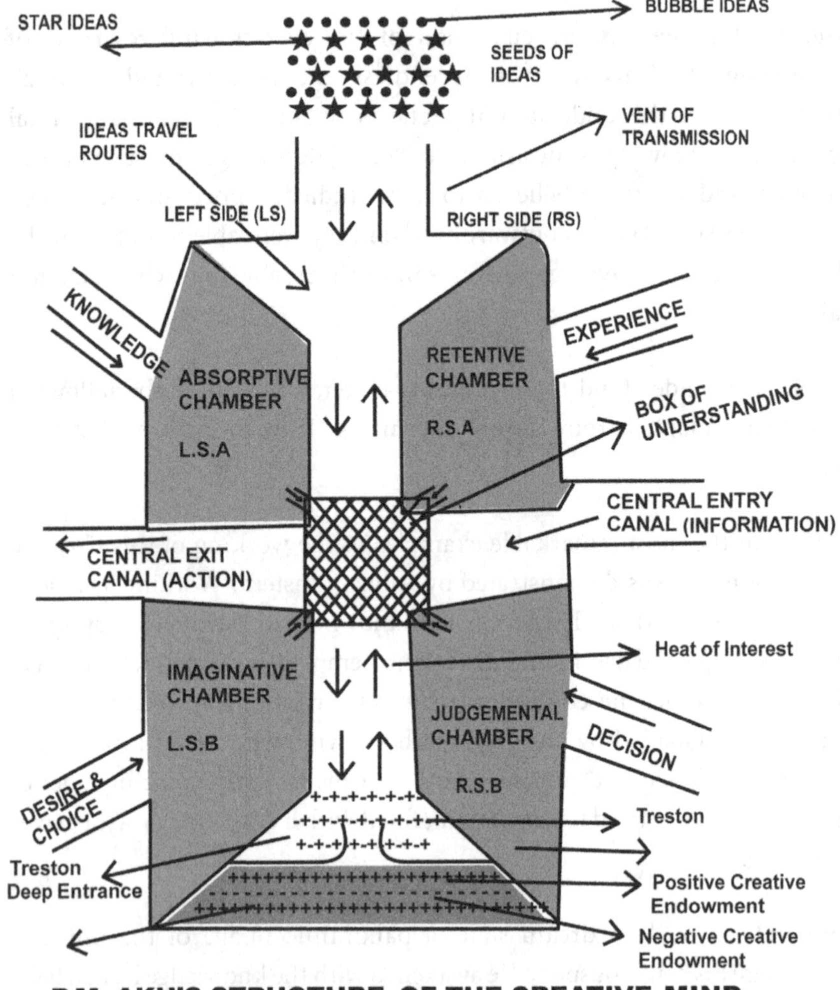

P.M. AKU'S STRUCTURE OF THE CREATIVE MIND

L.S.A. - Left Side Above
L.S.B. - Left Side Below
R.S.A. - Right Side Above
R.S.B. - Left Side Below.

We can't always be Einstein, but we all can,make much more effective use of our powers of creativity. It all depends upon the kinds of images we place in consciousness.

2. The story of the inventive genius Thomas A. Edison has often been told, how he had leavened to call upon his Creative power of mind, how he assembled all the facts he could determine about an object he wanted to invent, then turned all the experimental information over to his subconscious and retired to a cot in his laboratory and let this imaginative faculty got to work on it. In due course of time, his subconscious supplied him with the answers and ideas he sought. Edison recognized that his conscious mind contained no Creative power in itself - that it was only a collector of the experience and the information needed for the Creative power to produce the end result".

3. It must be clearly understood that this function of mind is not a form of so-called black magic or hypnotic influence... I have stated it conviction that a part of God, the Great Intelligence, dwells within each individual consciousness and that this constitutes a possible communication network".

It is however, important to be careful about the kind of information which we take into our mind. It should be intelligent and distinguishable from chaffs and loads of unnecessary mind-wrecking poodles which passes us daily. As described above, information creates ideas. The quality of a creative idea is determined by the quality of information which a person receives. Idea could either be a star (*) or a bubble (o) (Please see diagram). Get its entry into the rent of transmission is often dependent on the choice of a person. While a star idea has an in-built potential, the bubble idea is only a farce and has nothing valuable to offer. The study of the Creative mind guides to understanding the principles of a Creative Economy since it is persons who drives the economy.

RELATIONSHIP BETWEEN THE CM AND PASSION

Creative mind of a Creator to bring certain ideas into existence.

1. The CME is determined by the level of interesportil.

2. The CME determines the extent of passion developed for an issue or creative object.
3. If the CME is high passion is high; if the CME is low, passion is low.
4. Passion is the level of enthusiasm developed by the CME in order to create.
5. The combined statistical height of the CME and Creative Passion (CP) determines the amount of Prowess and the extent of creative talents applied to the intended work.
6. People with high C.P are often influenced by the CME to see and create more opportunities in every ventures they undertake.
7. The CME and C.P. Are both subject to statistical rise and fall.

GRAPHICAL ANALYSIS

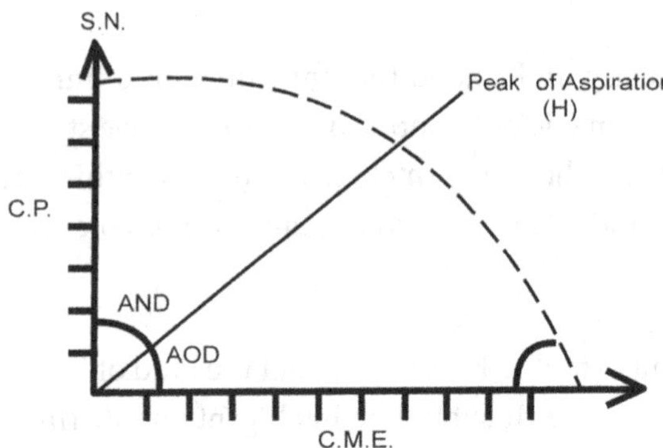

SN: Strategic North

The CME is directly proportional to the C.P. And equal to the height of creativity.

Mathematically expressed as: $\frac{\text{C.P.} = \text{H}}{\text{CME}}$

The CME forms the basis of the C.P. As its background, out of which there is no C.P. The augular rise and fall of the CME and CP prove that the TTL of any person does not follow a straight or head-on pattern since there is always the angle of decision (AOD) which may be started at any time. This process is what makes decision making difficult. The totality or sum of the

CP and CME must be equal in order to achieve a resource - able creative height.

12. PRINCIPLES OF A CREATIVE ECONOMY

    The P.M. Aku's 12 Principles of a Creative Economy are stated as principles and guiding laws.

1st Law: All minds in an economy are creatively endowed yet not all minds are actively creative, thereby causing a dispersion of economic activities e.g. Employment to unemployment; challenges of change and economic stability.

2nd Law: Every economy consists of resources, which have inmate potentials for functionality and development.

3rd Law: Every economy has functionality depending on its efficient use of resources.

4th Law: For an economy to become highly developed, it must create functional systems.

5th Law: Systems found in an economy depends not upon any-thing else, but appreciation of people.

6th Law: The people waking up an economy is referred to as the human resources, whose functionality determines the progress of a nation.

7th Law: A system is dead if it is not tranismissional i.e, does not posses the characteristics to be passed on to future generation usefully.

8th Law: The flow of cash without efficient system in multiplicity, soon leads to dryness.

9th Law: The strength t and functionality of an economic citizenry determines the extent of development with a view to creating a greater population.

10th Law: Service are product of functional systems strengthened by greater population, with regards to distribution.

11th Law: The efficient use of resources and its presence leads to society's functionality.

12th Law: The greatest resources presence in an economy is its human resources; for people posses natural endowments to enhance systems functionality.

From these laws, we deduce thus;
The mind is the Universal arbiter; so called because it decides the entire line of action of man in the Universal sequence of choices and consequence. Information received produced action given.

5 SEARCH FOR ORIGINALITY:

The Greeks were no pirates, nor were they vain imitators of works done by others. They believed in their own senses of creativity. Imitation. Of anyone's work or personality was abhorred. They appreciated the challenge and rewords which came by creating originals from previous originals. Some of the world's greatest movies such as the Trodgan Horse, Archives, Helen of Troy, have made lasting impressions Algebra, the Calculus, the Almighty formula, these are some of their classics.

Aspiring writers and creators must discover the secret of creating an original steam in the realization of a fault or deficiency which at the time of study or research existed without a common public notice.
Original can come from originals of previous works. Anything which falls short of originality loses its value from the beginning. It become artificial or imitation. The genius has a vision beyond the eyes of common place and that vision is to create a niche for inset by an originality. To obtain an original, the desire and search must go down to the Treston or bottom of the sub-conscious.

ACADEMIC QUALIFICATIONS:

In the world of academics, there are titles or qualifications which a graduate is decorated with, based on his performance 'His'; here, for the purpose of this book refers professionally to males and females alike. Before person can attain such qualifications. He must be tutored or must have passed certain standard test. I, personally do not believe in limits set by our conventional schools, either do I subscribe to standards which are lower, but I advocate through this medium, one must strive to exceed such limits, if notin school and examinations, there in performance. One must strive for excellence in a chosen field or career. Interestingly, most people who performance the most in life's creative place were those whose records were filled with examination failures, poverty, school dropouts, poor performers. Nobody was created for stupidity. While a school should never write. Off a child, the authorities should seek such child creative potentials in his various expressed abilities. The need for this lead to my discovery fo a program me called student career analysis program; which I predict shall one day turn every person who participate in it a creative genius on attainment of creative success. So much for the creative world. I believe in what? Shall call the creative world. What are its features? These include discovery of one's potential, self-retiance and employment, job security, job satisfaction, personal fulfilment, growth. Wealth creation: positive impact and influence, and good life. In the school of creativity, a student's early study of attendance would be:

Level 1: Interest
Level 2: Knowledge
Level 3: Research
Level 4: Discovery & Originality (D.O)
Level 5: Creative Success (C.S)

A personal who creates a <u>Masterpiece</u> would be awarded a master of creative sciences i.e. <u>M;CS</u> (male and mistresses of creative sciences and (mss,) for females. American Heritage Dictionary, third Edition is "An outstanding work of art or craft; something superlative of its kind". Such a person must be able to defend his work and express its originality, his motivation and his message.

Next, this person has great potentials for growth. I have discovered in history, a special respect for creator with at least five outstanding works and beyond. That would imply that the world expects more from a creator's mind. This form the next level of creative success attainment. An M;CS (see above) who goes on to attain five more works of outstanding feat should be awarded the title; Genius of Creative Sciences i.e G,CS (R) (male) and where mistress De Genius of creative sciences (MDG. CS) Female.

(R) means "Respectably": it is believed in my opinion that this person must have spent a minium of three serious years. His continuos effort must be assessed based on his motivations and the body or institute must continue to monitor such people, despite their graduation in order to offer more deceiving titles. These people can be trusted for certain important positions in society. They can be given special contracts to search their minds, within a special time on possible solution to impending problem of society such as employment, creativity science etc. This person is fit to teach, lecture and impart such knowledge to the others who are aspiring.

The third and the last stage of shared came from a careful study of people whose works broke national boundaries in greater numbers than five. They were recognized internationally for their over whelming works. Though some of them were young at the time of this recognition, their creative works spoke volumes,presently, you are perceived a creative success, even if one of your works sold globally or nationally: it should go beyond that of number of copies sold but number of works also. Interestingly I could sell a hit song, or write a river book: but it is necessary to prove the world:the community of people, the fans, the reading viewing and the monitoring public that I, am a creator and I can do the same thing over and over again, anytime, anywhere. The highest award to be given to any one in the school or world of creative success is what I callthe sage of creative success i.e.

SAGE OF CREATIVE SCIENCE (S; SC (L)), where (L)
Means "for life male and Mistress De Sage for female i.e MSD; SL). Such a person would have produced a minimum of ten works which are outstanding and recognizable. I recommend that these could be a combination of science for artistic, music and its composition and development strategies which are being worked out already or a separate format of independent faculty e.g. In arts;(ten plays), or novels etc..

The word " sage" is described by the Americans Heritage Dictionary third edition, as A venerated, wise person; while the word renewable ; which

is commanding respect, by virtue of age, dignity, or position: worthy of reverence, as by religious or historical association.
The age can build institutes of creativity in his area of competence and chosen careers. The sage is a consultant. He can offer valuable advice on certain issues of personal and group counseling, conferences seminars, motivational lectures, communications help-lines.His works speak for him than fifteen e works for the charted sage.
It is also noteworthy that awarded certificates must bear on them titles nature of the works and a message pointer of the issue discussed (books); depicted (autistic): recorded music master plan national issue _NI: or international issue Int / I) which could come to education, defence,

Terrorism, petroleum etc. Such works must have been copyrighted and its certificate copy be submitted to the professional body. The design of a certificate lies the discretion of the administering body. Creativity is a gold mine waiting for you who is daring to take up. The most prestigious tittle chartered sage is same for both gender.

What about the "What About".
"What about's are you talking about?. "I can hear you say. These are few bothering questions my mind and I believe should be on my mind also. These questions are:

1. Could a person be called an oracle?
Of course, yes see the definition of the word oracle. But I say that in this context: not in the evil traditional sense but intellectually or creatively; the sage could very well be referred to as oracle of creative science. He can be consulted and deserves his praise. So what about calling me (the author) an oracle or better still a sage?

2. What about the name given to an author's secret place of writing?
Form my research an doctor's operating room is a theatre; an acting place is also a theatre; I do recommend that the author's writing room or creative room be referred to also as a theatre. Why because the creator must be discreet during his commune with the higher and yet deeper powers of creativity; dissecting and knitting issues on scripts and walls.

3. What about Education?

Here I shall refer to what I call Educational Balance. This I shall divide into two kinds:

*. Functional Education: also known as practical educational, it bears its direct impact on society by its works. It depends on intelligence, rather than brilliance. Here creative success belongs. This develops skills basic for human living and occupation.

*. Theoretical Education: Also known as principle education. It depends on principles guiding issues. This is divided into (a). Active (b). Dead.

(A). Active principle are workable and are used in line with innovative creations e.g "skills make you rich not theories".

(B). Dead Principles: They are knowledge which are not really relevant to our present life. They includes undue arguments and promote crisis e.g man developed from monkeys.

Summary

1. Comprehensive personal research must be based on one's personal conception and deduction's as hell as in others.
2. The oracle refers to a superior source of knowledge.
3. Muse is believed to be the guiding power behind every creative inclination.
4. Literature is a written body of creative work.
5. Renaissance refers to the recognition of every new study or pattern of life with respect to the time of occurrence.
6. Theatre is any place where creative works are done extensively.
7. Classics refers to any work considered attractive.
8. Interest is defined as that point of catching the seemingly absent.
9. Formula for Interest may be expressed as:

$I = \frac{CTP}{OC}$ its unit is the CJ.

10. Interest exists in parts known as the Thermometric thinking level (TTL/T2L); which exists in the parts of low TTL (Grade I & II); medium TTL and High TTL, with categories measured in percentage.
11. Creative knowledge is the storage of necessary information.
12. Types of creative knowledge are (a) present (b)Absent.

13. Present knowledge are divided into (a) Present Usable Knowledge (PUK); (b) Present Usable Knowledge (Puk)'; (b) Present non-usable Knowledge (Pnuk).
14. Absent knowledge is divided into (a) Absent Intuitive knowledge (AIK); Absent Uncommon Knowledge (AUK).
15. Pmakumatically Knowledge is expressed as:

$$KTP = \frac{Kn \times Ex}{De + Ab.}$$

16. Creative research consist of (a)Oral (b)Intellectual (c)Personal
17. The use of the mind creativity is necessary for every creative ability; which leads to the principles of a creative economy.
18. The mind is the universal arbiter so called because it decides the entire line of action of a person.
19. Originality characterizes true creativity.
20. Creative sciences is academically qualified by the (a) Master/Mistress; (b) Genius/Mistress De Genius (R); (C) Sage/Mistress De sage L (d) Chartered Sage.

CHAPTER 6

P. M. AKU'S 60 PHILOSOPHIES OF CREATIVITY

Aims / Objectives:

1. To prove the large extent of creative considerations available.
2. To show that issues which relate to creativity are practical in every day life.
3. To prove that creative endowments are not myths but provable facts and physical evidences.
4. To teach the most modern applicable philosophies of creative abilities to careers, individualism and development of society.
5. To show the relationships in referenece to differences, uniqueness and similarities in creative matters.
6. To show that creative endowments are the very core of existence.
7. To show by applicable philosophical thoughts, that every society is either wealthy or poor depending on the extent of emphasis placed on creative contributions of individual citizens.

P. M. AKU'S 60 PHILOSOPHIES OF CREATIVITY

PHILOSOPHY 1:

Discovery of a talent or special ability may be called a gift because it is specially endowed upon such as person.

In the issues of life, the first step towards successful living and fulfillment is the discovery of a person's purpose. This is usually expressed in such a way that no other person can do so except that person.

PHILOSOPHY 2:

A person may develop a talent to such an extent as it is never imagined, or done ever before; then he becomes a master; a genius; a sage; depending upon the strength of works and thereafter a hero, a role model.

This is true and proved in the fact that every talent so possessed and discovered must be trained, guided and developed. When a child draws or carves, a beautiful object or scenery, we admire it, it appeals to our senses and we praise it; yet our greatest responsibility lies in providing the opportunities for the further development of such an expression. While an original work may be referred to as a masterpiece, he must be developed to become a creative genius, a sage or a chartered sage thereby growing to become an authority or role model in such talent.

PHILOSOPHY 3:

A Talented person must be docile enough to learn new experiences; understanding enough to reason over it; intelligent enough to find a place for it in the ordinary affairs of life and courageous enough to defend it.

In the pages of history, very few people out of every generation have been termed great. This is because must people become overtly excited when they learn the rudiments either by research or proper training. Docility simply implies humility, understanding comes often by careful study, intelligence is obtain by wisdom and considerations with reverence to God and respect for human dignity and the courage of expression in the last ingredient of an endowment. These can not be complete if anyone is lost. Creative talents; when discovered is like a flame, it can burn and it can warm; it is a risk in itself but must be calculated to obtain its very benefits.

PHILOSOPHY 4:

A person may be trained in a field of his own liking; may be adventures in such that thrills him; yet may never find a purpose except in his talent.

I shall tell my personal experience. Due to my love for adventure, I had a training in Aviation; I was one-time student of Law, business Administration and attempted entries into the Nigerian Defence academy without success; yet I had my eyes on doing something extra-ordinary; I wanted to break a world record, I wanted to be different, I wanted to become a strategist; though no special school taught a course on strategy; I hated the news of unemployment, poverty and the inability of people to develop themselves. I simply saw myself as a hero and my guest was much more than financial security; I wanted to find my seat in life and sit on it. To everyone, I was a bundle of problems. Yes, that's right, I discovered in the biographies of great men and great societies, that great deed were once a bundle of nothings;

greatness itself began within a person from the spirit of confusion. Today, I have came to learn appreciatively that many empty seats for greatness are yet to be filled. Creativity, I search out these vacancies and hope to fill as many as possible. Remember, it's a choice.

PHILOSOPHY 5:

There is such thing as practicality of a field by training and such by calling.

Now, when creative potentials express themselves initially, it does so as unexpected bumps waiting to be harnessed and smoothened out. If we just climb over, you lose that aspect of it, then there tends to be no rest or internal peace until it is addressed. The logical essence of education is to train the mind to be liberal enough to face life intelligently, solving problems and creating a positive impact to humanity without fear, contempt or superstition. Our educational training ought to develop our discovered potentials. Once upon a time, there was no gravity in physics until Sir Isaac Newton at barely 23 years discovered it and then the three laws of motion and then a course called calculus. These laws became the ideological revolution to the creation of automobiles, aero planes and rockets. You see, you're important.

PHILOSOPHY 6:

Every true talent comes with its own original blue print; every artificial of it is a conspiracy against nature.

Originality is the strength of creative endowments. It goes beyond mere imitation. Creative conspiracy occurs when people steal, or maneuver another's original idea to earn from the benefits or proceeds. One thing is certain in such a situation: that it will not last long. It is the creator who understands the longevity process of his original idea or concept. He who steals another's idea cheats on his own endowment since he cannot find nor develop them. Thank God for copyright laws.

PHILOSOPHY 7:

Creativity, I foresee, is the last hope of mankind; this can only be found through a person's talent discovery; one whose ultimate purpose is to glorify the creator.

The essence of humanity is service, to God and his fellow man. True service esteems the productive use of creative potentials. The world is enjoyed by heroes, yet everyone is a potential hero. A nation which values her people does so by the persistent and individual contribution made through creativity. As jobs are lost everyday, families breakup and youths degenerate, the government will do well to encourage creativity and its studies in order to regenerate, rejuvenate and reinvigorate the ailing economic system. Creativity help train the necessary habits of hard work, patience, hope and love thereby ensuring true happiness. It is the last hope of mankind's survival beyond every dispute.

PHILOSOPHY 8:

The earth has been created with certain opportunities for innovations, so man is hereby qualified to beautify and maintain it; not to destroy it.

Opportunities for greatness often present themselves in the disguise of difficulties. Our own requirement would be to look at ways of harnessing its advantage. He who has no job should either search for one or create one. Creative potentials have ways of attracting people of like minds. Nobody wants to invest in an empty brain. Nobody is a friend to a nobody. It's not about wealth which may vanish, it's about creating a legacy out of your potentials in order to keep your financial security intact. Not even heaven praises the waster or the complainant; but now and then those who creatively solve problems with the available resources tend to become noticed.

PHILOSOPHY 9:

Every man is important to the other; no profession is more noble than another.

Another term for this philosophy is what I term the cycle of professions. If there were no teachers, there would be no education; no lawyers would cause chaos and anarchy, no pilot driver, men would be stagnant; no preacher, then loss of morals, no farmer no food; no government, no governance; no engineer no contraction; no doctor implies no cure, no scientist would lead to no discovery and prescriptions to the doctor; no writer no reader and thereby less intelligent information; no hairdresser, no hairdo; no businessman to sell the goods and services, perhaps men would chew their products. You see, everyone is important to the other.

PHILOSOPHY 10:

When a talent is turned to a purposeful use, it becomes a career, a business, a world class venture, because someone was courageous enough to strive research and skills development.

Interestingly, every thing we see and use today is actually a product of an idea. The strength of this may be proved, for instance, in the brand models of products e.g. cars, which were used in the 1970's, 1980's and 1990's. Today, new models for to narrow are being produced. Every creative endowment is a sitting industry when properly harnessed.

PHILOSOPHY 11:

I am talented in a field or skill and doing nothing about it; is a proof that something is missing; that I have not properly done my home work.

Every collective progress is the result of individual contribution. A society fails when individuals fail to develop their uniqueness. This would lead to over-dependence, insecurity, high crime rate and social disintegration. Whether encouraged or not, the individual must consider himself duty-bound to make his positive contribute to the society. Hence, those who have performed excellently are those honoured. Then would poverty and mediocrity be a thing of the past.

PHILOSOPHY 12:

Talent harnessing is no wild goose chase; human development is no futile dream; it is a worthwhile career.

Every great career started with a simple idea; concept development and overall impact on humanity especially where and when it was deemed necessary. The essence of this work is to develop the talents within through certain applicable principles, without suffering undue mental stress if studied at the initial general course level in higher institutions; and to bread a new generation of creative scientists when studied at the professional stage; all proved by works and without bias or questionability. Rather than picking just a person from a crowd, I recommended that programs developed should be geared towards overall development of participants in order to encourage greater efforts in everyone. When skills are developed, purpose is meaningfully driven and life is much more fulfilling. If you wish to develop your specific skills, study role models; and if you find a soul willing to imbibe yours, then it's your own opportunity to earn through impartation.

PHILOSOPHY 13:

The person, be he youth or adult, who denies himself the truth of his talent; in pursuit of money for a worthwhile gratification, denies himself, his own reserved seat in life.

Money is good but is not every thing. Money is useful but guarantees no security when it does flow from a creative reservoir. It is necessary that we develop on potentials creatively, though working to earn livelihood for temporary usefulness. Everything considered worthwhile would not have been if their inventors or authors sat down all their life waiting for a salary increase. You must use the previous experience grate fully to create and control your own destiny. Like Abraham Lincon once said, "I shall study and wait and my opportunity will come" we do know his story to American House as President. Come to think of it, some of Ameriais' greatest presidents were also respected as highly creative persons every before they were compelled to mount power. People such as Abraham Lincon, Thomas Jefferson, Benjamen Franklin, John F. Kennedy and persons like Henry Ford, Albert Einstein, Thomas Edison influenced the policies of the American government as creative strategists. Then Ralph Waldo Emerson influenced literature along long fellow and others, while John D. Rockefeller discovered economy and entrepreneurial creativity in developing new business modules, while being practically objective. Choose where you belong, there's still room for positivism because the top is never crowded.

PHILOSOPHY 14:

As long as people are directed to remain employees of another person's discovery, with the fear to dare upon their own lots; there may be more of economic casualties and job insecurity.

Studies and experience have proved that long service as an employee tends to produce negative effects on the mind of an individual.

Robert Schuller, author of "Tough Times Never Last, but Tough People Do" calls it "Locked In Thinking"; Robert Kiyosaki, author of "The Business School" calls it the "E quadrant, E stands for employee… In other words, security is a very important core value for someone in the E quadrant" I shall call it "Mind Slavery", or the "Employee Pressure Syndrome", the "Millenniums Bug", the "creative Sabotage". Working for another in itself is not bad, please don't get me wrong, but working for twenty, thirty, forty years with an eye on a certain amount of gratuity or pension only hampers your explorative powers' opportunities, then the government is guided; but when they don't, then everyone, the system itself suffers.

PHILOSOPHY 15:

The person who tells himself truth ought be praised as the most loyal of all men.

Creative success never occurs by chance or coincidence, it is a product of honest consideration of one's abilities, resources, risen and realistic view of a person further, we shall discuss the philosophy of raydealism, the creative person attitude. When a person tells himself the truth, certain realizations are made, opportunities beyond the present state of things are made dearer to him Francis Bacon, a great philosopher and reputable Essayist once said: "Lean freedom is better than fat slavery". The battle against individual and economic sabotage is either done now and won, or we continue to suffer and waste. It is about me, and it. Is about you; now or never.

PHILOSOPHY 16:

Talents have brought more glories to people, opened more ways to the big gates of high achievements; directed lights to the once-upon-a time-nothings; and reserved fame for their posterity than any other

Imagine being related or married, or better still affiliated to a great musician, inventor, writer, business. Strategist, teacher, scientist, lawyer, banker, or captain of industry. How would you feel? Elated! Happy! On top of the world! That's what's what I am talking about there are people whose wanes open big gates; whose autographs or signatures of them either inspires you or demeans you. these are people of creative wonders, dead or alive who teach the world what to do and help it practically rolling along, you can be there also.

PHILOSOPHY 17:

The beautiful ones are not yet born; the best designs are yet to come; and we are yet to behold the greatest and the greatness of young ones who know and live upon the practicality of an education based on talents and creativity.

This is a call for a revolution of the present education system, an education based on functionality can only be obtained by the teaching of creativity and its guiding principles. This would revolutionalize the dull system present in today's research poor reading culture, and adherence to the rules. When this kind of education is encouraged, then the sky would no longer be a limit but a starting point.

PHILOSOPHY 18:

The greatest of all things are those not yet seen; so is inspiration-the father to creativity and a companion to confidence.

The Holy strictures says in the book of proverb 29:18, "Where there is no vision, the people perish: but he that keepeth the law, happy is he. There is no vision with out inspiration. When people are inspired, they tend to foresee the future away from their present state.

This, then, breeds the development of the seeds of developmental strategy in them. They become moving loads of opportunities and world changers.

PHILOSOPHY 19:

Inspiration is no accident; talent is also no coincidence; for where you find one, other accompanies it.

According to Thomas Edison, "Success is 1% inspiration and. 99% perspiration" when minds are creativity dead, they cannot be inspired. Only motivated minds can be inspired; and only purposeful ones moves to action to bring through creating the hidden substance within their own minds.

PHILOSOPHY 20:

Why marvel at this; those linings upon your palms share no single room nor likeness to that of any other in the entire Universe; your own computer chips they are and the structure, only your mind can teach you.

There is a great intelligence behind our very existence. We didn't just come to be. We are no substance of the misleading theories of evolution. We have within us the spirit of God our creator. Our creative endowments are the sparks of God's creative variety. Individually we are unique, but in Him, we are complete. Our faces are different, hardly do we see a set of twins with complete the same genetic make-up. Moreso, our finger prints and palm linings are different from each other's. this is interesting. I am unique. I am special. In the entire universe, there is only one me and after words, there can never be another me. The Bible speaks of David's confiscation of this philosophy in Psalms 139:14; "I will praise thee: for I am fear fully and wonderfully made; marvelous are thy works; and that my soul knoweth right well: God expects us to continue where He stopped and give him the reports of our deeds.

PHILOSOPHY 21:

Believe in your uniqueness, your specialty; nothing good is odd, you are a fraud if you dance against the heartbeat own gang.

He who refuses to rule his life creatively will have others run it destructively. Believe in yourself. Believe in that good purpose God has laid within you. If you try to express it in one way and it fails, try another way. But believe it enough to work it, to express it, to win for it, to live for it and be willing to die for it, then you'll win. Outside this, you may never know the joy of fulfillment. It makes no sense living a routine life throughout life without having and purpose. If you do not believe enough in yourself, remember the people do: God your creator and myself. So don't give up. Go! Go! Go Now! Start!

PHILOSOPHY 22:

The Maxim "Catch them young" remains the most indelible discovery of the human race because a person is a product of his own training and mindset.

This informs the special reason why I have special interests in youths and skills development. The children of today are the potential leaders of tomorrow. I have used the word 'potential' because only those children properly trained for leadership will eventually get there. The world is dynamic and change is gearing towards creative leadership. Home training and domestic skills is necessary in our present age. I call this age the Info-Time age fast approaching from 2010, an age or period when information available would inspire children at early age to work ahead of time; perhaps the "continental clock positive system" (CCPS) would be a better term for it.

If the youths must be the custodians or posterity's keys, and use it efficiently, then I recommend special training platforms for studies in creativity. For instance, this inspired me to create the students Career Analysis Program with a section for Junior, Senior and Adult categories. They must be told the truth if they must grow about the standing realities of life while they are helped to cross the hurdle before they get there. Perhaps, if we had been taught on time, our youths wouldn't be suffering harsh economic realities today. This is the opportunity for the resourceful.

PHILOSOPHY 23:

The memory of an adult is after the product of his childhood activities.

If the mind as proved by psychologists is a custodian of records, their this philosophy is true records once had it that the great king of POP, Michael Jackson was tutored under his strict father. Often his dad would scare him wearing masks to remind him that he ought to close his windows at nights. From my analysis, what is the after wath effect of this in his adulthood was the transformation of Michael Jackson into a wild beast in the musical Video, "Black or white", the ere scary transforming beast and the dancing zombies at full moon (midnight houor) in the video," Thriller," The creative mind of this man simply helped him to put that experience into an advantageous use what did that earn him; Grethess! You must search for every necessary experience of yours to put to good use helping to transform lives positively. Cases of child abuse, emotional in balance, poor grades of school days, adventurous movie etc should be used on the positive note to re-orientate our society.

The memory is a great supplier of stored or reserved knowledge to the creative mind. it's resources, good or bad, must be totally recreated to remodel society, for better.

PHILOSOPHY 24:

Education means to impart knowledge. wisdom means to make productive use of the imparted knowledge based on experience; but without discovery, any of these would be a waste of such an opportunity.

If education means to impart knowledge, then you must permit me to ask the question; why are the educated unemployed? Why do the uneducated tend to be better fixed earlier that the educated? Allow years in school studying could be regarded as a careless waste, is that true? The answer is in the type ox education we obtain. We simply obtain more theories, less skills kind of education. We are full facts which do not give us any due to how to solve our problems. Our researches are not updated enough. Interestingly, History is full of persons who broke out of school and ended up as heroes, revered and respected. What was their common secrets? Cheek it out! They were all creative. They learnt skills by obtaining and updating on certain knowledge. They understood the precepts of knowledge and wisdom. That is right. We need an education based on wisdom. This can only be creative education. Since creativity is all encompassing, then it is wise that it is made a general studies course or subject for every one. Great opportunities are an evidence of great discoveries. Great opportunities are an evidence of great discoveries Great discoveries are found in creative education. This, permit we to say, is much more than teaching rotational studies, but developing the mind to find and nurture it's own bearing. Thomas Edison was not the first to work on the electric bulb, yet he perfected it and practiced its marketing and selling skill. This is a classical reference of way others. I think it's time we start thinking.

PHILOSOPHY 25:

Discover the person; tell him who and what he can be; and you strike against societal poverty by one blow; then tell the same to the same to ten others; then a thousand souls are saved.

The best way to win souls is not the Hearen-Hell scare. It is not even threat of punishment. It is the careful development of the wind positively. When people are taught that they can succeed by believing in God and in their creative powers and abilities, their response tend to reciprocate, and their zeal fired. You feel the passion within them burn for achievement. The principles of creativity, when taught fires the mind of imaginative process there by enabling the 'ICON' spirit and attitude in such a child. This philosophy worked for me while I practiced as a teacher. Now, it is still working in the career analysis program. Emphasis should be laid on developing the mind first, then the positivism, he tends to spread it to others because emotion is a beat, passion is a fire. So, everyone is connected to someone. A person who has learnt to benefit himself would willingly accept to benefit others. He would consider it a product of integrity to touch other lives positively through his works and purpose. He would work for the gold not the pory packet since riches would naturally come to him. This is true beyond every doubt.

PHILOSOPHY 26:

As a matter of national education policy, let every graduating person teach what he has been taught; what he knows, and then he would wish to practice based on learning more; for as you teach you also learn.

There has never been a great person who could not teach what he learnt and knows concerning an issue. History confirms that people are great when they have made a great thing and have gone ahead to impart same to others. Unfortunately, people graduate with an attitude of pride and arrogance. They move about wt an air "I'm above all", proving the fact that they do not understand that greater responsibilities already lays hold on them. Then in the face of economic realities, they are drowned in the myrah of stupidity, smoking and drying in the part of poverty and than gradually they join the withering crond of confusion letting depression, frustration, rejection and complain takeover their lives. As usual, their high hopes are disappointed and their language become complain and blame against the society, accompanied by sighing against the government for not creating the jobs that they ought to create by the study of a course in the higher institutions. A creative person must be a teacher because when his works attracts criticisms, then he must take corrections, and when he falls, it's time then for him to rise. By teaching, ideas are exchanged and strategies are developed, so that by teaching, he is also learning and building the necessary confidence to face the approaching future.

PHILOSOPHY 27:

The secret of discovery lies in wearing a human face and saving your heart for empathy.

There is tremendous power in discovery. In order to discover, you must pretend to yourself that you know nothing. By this, your creative mind is ready to learn and purpose something new. My purpose everyday is to learn something new and worthwhile and this forms the bedrock of my prayers. Then I view everyone as a teacher whom I can learn from, as this helps me to listen; then I also view each person as a potential I can tap from. What

I get is eventually added to what I do already know, helping my intellect to expand the more. And moreso, books!.

PHILOSOPHY 28:

The amassing of fame and wealth lies in offering a worthwhile price for a life long discovery for a greater people who see it as opportunistic.

This is simple business strategy for success. Create and sell with the average person in mind, at an affordable price and then you are made. When people view a work as cheap, they also tend to see it as opportunistic, thus making them willing to buy. Creativity entrances the observation and application of simple strategies to create your own place in business. It enhances competition, high quality goods and services, productivity and better standards of living.

PHILOSOPHY 29:

The youth may seem unimportant but not so much as not to influence his parents' pockets.

This simply holds the lesson that parents wish to pay for the happiness of their children. There have been situations where a child recommended his teacher to his parents. And what followed? A closer relationship which brought enormous blessings to the life of such a teacher. The creative person must learn to respect the child too.

PHILOSOPHY 30:

Inspire a youth for greatness, he becomes your loyalist, empower him; and his parents shall remain at your service.

Creativity does not hide. A child by his innocence could become inspired by your works, then there is a relationship direct between you and indirectly connected to his parents. It is the benefit of creativity to help others too.

PHILOSOPHY 31:

Nothing is more bothersome as an untrained youth; nothing more troublesome like a decadent one.

When a child has defied every reasonable form of discipline, then try finding out his interests that is his creative abilities. By this you would know his mind. Work on that mind. At youth, he struggles with the vision of the future, parental hiccups, peer pressure, puberty and adolescence and you need to help him develop the right mind in facing these challenges.

PHILOSOPHY 32:

A people; a nation waste at a time for every wasting youth; but succeeds a hundred times over for each discovered and developed one.

This hinges on the kind of education given to them and the general attitude of the people towards developing creatively.

PHILOSOPHY 33:

The essence of academic study is to foster intellectual research; the essence of creative studies is for the enhancement of talents and skills; job creation and security; self-employment, proper application of intellectual research based on basic evaluation of practical experiences from daily practices.

PHILOSOPHY 34:

The more we engage in a thing; the greater the chances of understanding the process.

That's right. Understanding the process requires that you get involved, carefully observing and creatively studying its operations; nothing faults and innovative options and opportunities available. That's how great men have made great things in all works of life.

PHILOSOPHY 35:

Diversification of talents arises from the simple understanding of basic principles and concepts of other aspects and things.

Creativity is a central junction, a point to many routes. You can choose to know one and stop there; or you may travel each route one at a time and yet get to know them all.

PHILOSOPHY 36:

As long as there are problems, wealth comes from the creation of its solutions.

The creative person does not run from problems. Rather he is fulfilled by each problem he can creatively conquer. That should be our national spirit.

PHILOSOPHY 37:

The mind is not limited to singleness, but it is open to variety of interests. This is proved from the fact that a person may wish to work on the course of his training and simultaneously on his specific identified potentials.

PHILOSOPHY 38:

A person through resourcefulness and diligence, can become professionalized in many things by simple understanding and practice. The most worthy goal of the creative mind is to enlarge his capacity to grapple with many things. However, the conditions to attain professionalism in these lie in resourcefulness, diligence, understanding and practice. History has it that Lord Dening, a Justice of the United States Supreme Court once felt cheated that he did not know much on a science course, he had to enroll for this course, and despite being a lawyer, he graduated in that course a success. The level of your mind development will determine your innate ability to grapple successfully with many pies. This philosophy justifies my ability to cope with being the World's 1st Creative Scientist, a propoplay wright or Peterwrit, a Drapoe, a playwright, a poet, a novelist, a consultant, a teacher, a strategist, a script writer and then a Christian and father. You see!

PHILOSOPHY 39:

Taking the initiative is a difficult process; but once taken, it becomes easier to live with it.

One of the greatest intellectual disaster of the lay man in our days is that he doesn't want to know and he would arrogantly boast that he doesn't want to know. According to William Shakespeare, "Folly and ignorance are man's greatest enemies". Creative success does not occur by magic. It occurs every time an intellectual enquiring is made into an issue to determine why and what is good. Taking the initiative to solve an existing problem is a challenge to every person which brings many rewards when surmounted.

PHILOSOPHY 40:

Education in all is encompassing; creative education inevitable; while agriculture is all important because a hungry man is liable to suffer intellectual frustration.

Through my developed Career Analysis Program I have been able to introduce certain polices. Firstly is the compulsory study of geography since it touches on every field aspect e.g. Geology in Science, Cultural in Thropology, History in Arts; and general business in the commercial department of study. Also a dictionary of careers to help students understand various fields of study and vocations as recommended, an annual year book to encourage the reading culture, agriculture study for all science students. These and others practically took me to the heights in intellectual development which I have attained. I do sincerely believe it would do the same to our youths of school age and adults as well.

PHILOSOPHY 41:

Every process involves a study of some sort, a science; so is human behaviour, talent and endowments.

This has informed the reason for this research on the principles of creative sciences. It is worthy of note that this course is not restricted to science students simply because it bears the word "sciences" attached to it. After all, the word science simply means knowledge (a Greek word: Scientia). Creativity is the essence of humanity and pure service to God. From my experiences, I have come to observe that when a problem is attended to accordingly with the tenets of the creative principles. Solutions come with guided directions. This is practical as you would find in the structure of the creative mind. It is no mystery, it's a careful study.

PHILOSOPHY 42:

Although all talents in a person are unique, there are certain relative joints among them.

In the creative mind, nothing exists or works in isolation. Results are a product of combined factors of why, processing and exist for profitable use. This is also justified in what I have discussed in the topic "Universal Relationships". A classical example of this proves when you study the linings on each of your palms. Each is distinct, yet relatively joined to another. Knowledge is complete when its facts are contributory. You can find a bit of Physics in Chemistry, or Biology in Chemistry or English Language in other subjects. So it is.

PHILOSOPHY 43:

While some talents in a person are dominant, others are recessive; yet there is the tendency to excel at more than one field or endeavour.

This will be dictated by your choice and level of practicable interest in a potential. When certain talents are fully developed to use, it shows where you have given your mind, energy, resources and time. The recessive talents are a reservoir of new aspirations waiting to be tapped. That's why you are here: to explore, discover, and use gainfully.

PHILOSOPHY 44:

The extent of light which a person would receive in life would greatly depend on the person's decision to work it out.

Decision; the king of every force seen and unseen lies waiting at your service right there in your will. You say go there, and he goes, stay there and he stays. That's the power given to every person by our creator to make choices whether good or bad. Our decision brings along its own consequences so that they are openly weighed ever before we make one. What you have done or failed to do in the past must have led to where you are today; and fresh decisions made today will also go a long way to affect your future. Son, it is imperative to train your conscience to help you make the right decisions every where and every time.

PHILOSOPHY 45:

A talent which cannot express itself in varieties is a waste of nature's creational resources.

After a thing has been made, how does one keep going? That is where style and innovations must come in. the creative mind searches all the time for new strategies, methods and productive ways of doing things. He is always ahead of time, projecting resources for use over a short period of a hundred years. If you have not got to this level, then you have a lot to learn. Decide which of these statements you would reply if nature questioned you on how you used her resources "Useful and multiplied", or "Wasteful and Fallen".

PHILOSOPHY 46:

There is mote to the wells of the reservoirs stored in the human mind than meets the eye; these talents or gifts must be searched, dug, discovered, developed and used in order to appreciate the mind.

Well, the taste of the pudding lies in the eating.

PHILOSOPHY 47:

For every evil, there's a counter of many good; in thoughts and deed; it all depends on our real use.

What side do you belong? The world has come of age to give you back exactly the rewards of your input.

PHILOSOPHY 48:

Talents are independent of gender, age, tribe, qualifications, position, social status, race and nationality; and language; and so a creative person may work anywhere appropriate.

I believe this soothing to everyone. There is really no excuse why we should not succeed in fulfilling our purpose. Education, however, should not be mocked and basic training should not be neglected. No one was created to be a waste.

PHILOSOPHY 49:

Talents do not die; but may be suppressed.

Those who begin their search early also reap the benefits early.

PHILOSOPHY 50:

The most powerful factor upon the human mind is information; powerful enough to stimulate, to motivate, to create, to save and to destroy.

Depending on the kind of information you obtain. Encouraging or discouraging; constructive or destructive, purposeful or purposeless, guiding or distracting; up building or down pulling; there's nothing welcoming as a good news; and nothing detestable as a bad one. The kind of information you obtain has a profound influence on your mind.

PHILOSOPHY 51:

Supreme knowledge rests in the hands of him who has discovered self, who knows where he is going, who has the unfailing determination for reaching the object of his heart's desire, who has formulated a workable plan for his attainment. It's in the individual.

PHILOSOPHY 52:

It is proved globally that only the citizens of a country hold the will to drive the wheel of their country forward. I believe the only way to achieve this is to operate the educational system as a collective responsibility.

It is proved globally that only the citizens of a country hold the will to drive the wheel of their country forward. I believe the only way to achieve this is to operate the educational system as a collective responsibility.

The achievement of worthy goals lie in the education system of a people. Education is massive in its operations and here both the private and public sectors must collectively develop it. It's our first responsibility.

PHILOSOPHY 53:

All other systems seen to introduce innovations continually. Yet our educational system ought to enjoy this facility first since it is the tree which houses the fruits.

Better systems of learning and better standards. All these are necessary tools for educational proficiency.

PHILOSOPHY 54:

Where there is lack of scholarship facilities, I doubt if the best brains would ever be discovered, since only the rich would be able to fund education.

PHILOSOPHY 55:

The fathers of our national freedom were products of educational opportunities; why should we not seek for more of such good opportunities?

PHILOSOPHY 56:

Sustain functional education; consisting few theories and more practice; then breed a nation of practical and sincere leaders.

PHILOSOPHY 57:

There is nothing you offer for the promotion of intellectual culture that is enough, it remains everyone's highest duty for time attainment.

PHILOSOPHY 58:

He who fights for educational emancipation must hold the law firmly.

PHILOSOPHY 59:

No education, no nation; no functional education; no functional nation.

PHILOSOPHY 60:

Education must lead the pace for others to follow; otherwise it would be the case of blind leaders leading a blind people.

SUMMARY

1. Only creative studies in its real essence can develop the mind productively.

2. Creative study is the mother of all studies, and knowledge ever bourse or conceived in the mind of a person.

3. The promotion of intellectual culture is complete by the appreciation of creative knowledge.

4. Creativity is a fugitive without study; a wasted knowledge without practice.

5. Where creativity is absent; destructibility out of ignorance takes present seat since there is no vacuum in nature.

CHAPTER 8

MIND MECHANICS & ENGINEERING

Aims/objectives:

1. To prove that the mind is an operative mechanism subject to analysis, development and improvement.
2. To expose the student to the often neglected parts of our mind which is central to all existence.

MIND MECHANICS & ENGINEERING

The Word "mind" is defined in the following ways:

1. The human consciousness that originates in the brain and is manifested especially in thought, perception, emotion, will, memory, and imagination.
2. Intelligence; intellect.
3. A person of great mental ability.
4. Memory Recollection.
5. Opinion or sentiment.
6. Sanity- to obey; to attend to; heed; to be careful about; to care or be concerned about; to object (to); dislike.

Characteristics of the mind

1. The mind can be filled up and can also be empty at the same time; halting much concerns about certain issues and being empty or knowing little about others. This is called **Mind Flexibility.**
2. The mind can exert itself to produce certain hallucinatory effects. This is the process called ***Mind- blowing.***
3. The mind has the innate ability to experience and take in situations and experiences which are perceived to be intellectually and emotionally overwhelming. This is called **Mind- boggling.**

4. The mind can be psychedelic - having the error developing behavior.
5. The mind can be expanded to understand further. This is called **Mind-expanding.**
6. The mind can be lowered to narrowness of intellectual capacity. This is called **Mind-narrowness.**
7. The mind when not propel cared about has the ability to lack intelligence or sense leading to carelessness and headlessness. This is called **Mindlessness.**
8. The mind can take complete care through being attentive and heedful, then it is said to be **Mindful.**
9. The mind can be trained to a particular understanding, mentality and attitude fixed that determines one's responses to and interpretations of situations. This is called the **Mindset.**

Though for the purpose of this study, we are not, mostly concerned about the general understanding of the mind or psyche based on the views of complete psychology where we find concerns in other mind various issues such as, psychiatry, psycho active, psycho analysis, psycho drama, psycho genic, psycho metrics etc. We are more concerned about the creative mind, its mechanics and engineering. Let us define further:

The word "Mechanics" in relation to this study is the functional and technical aspects of an activity e.g. The mechanics of football; while its connecting word "Mechanism" is a system of parts that operate or interact like those of a machine; a means or process by which something is done or comes into being; while being guided by the doctrine that all natural phenomena are explicable by material causes and mechanical principles. "Engineering" therefore is the application of scientific principles to practical ends, as the design, manufacture, and operation of structures and machines.

By these definitions; Mind Mechanics and Engineering may be defined as the understanding through analytical study of the structure of the mind with the view to manage it properly and direct it towards creative and purposeful ends. By the study of the Pitolemaic System, originated by Ptolemy, the Alexandrian (Egypt) astronomer, mathematician and geographer in Fl 2nd Cent. A.D.; Its astronomical system was that earth is at the center of the universe. This is applicative only when compared to the knowledge of our present day which has proved the existence of over 2,000 planets, the spherical nature of planets (including earth), the size of stars and the mystery of their illumination, and the sun as strategically placed at the centre of the universe, serving many purposes of helping rotation and revolution of the nine planets in the milky way or nebula. However, our study here asserts that the universe is bombarded by existential and living laws which are controlled by the mind of man. This proves that, without any doubt, and assertively so, the mind (which controls the earth as God's intelligence controls man) is the center of the universe. Wrongly perceived? Why then do we need to serve God, our maker, and nothing else? This is because they develop our godly capacity to create, develop and grow. Why do we need to believe, have faith, develop confidence, sincerity, integrity; why do we have to imagine, receive inspiration, see a vision and conquer the fear associated with death. This is because God's power is in us.

Proving this is the Central Universal process (CUP)

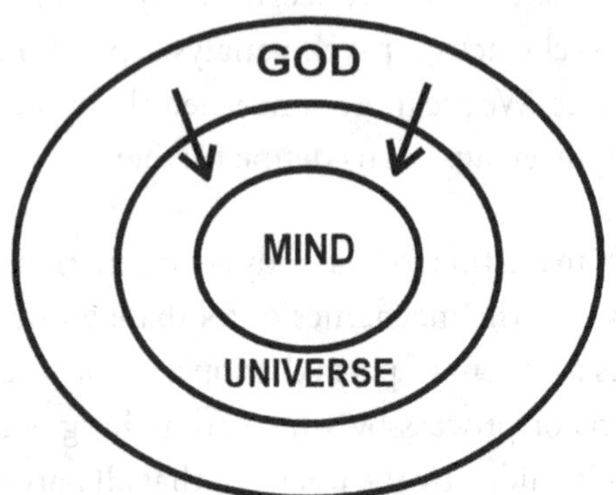

CENTRAL UNIVERSAL PROCESS (CUP)

Central Universal Process

God controls the universe. He has also put and endowed man with certain aspects and power to control the universe. This can be achieved by a total submission of our individual power to God willing in order for Him to direct it on what appropriate laws, to be applied for the universe to obey; what resources to be tapped and its efficient and the general control.

Discovered in this process are certain laws. This is what I refer to as **Universal Laws Of Consistency.**

LAW 1:

There's no vacuum in nature's universe:

Everything in the universe fills a space of responsibility and every space is occupied; or yet to be occupied by its occupier.

LAW 2:

Everything in the universe has a purpose:

The stars for the shine, the sun for the lights, the plants for the heights, and man with his mind; all things, everything works together in the universe.

LAW 3:

Everything in the Universe exists for the other and thrives in a relationship for the other.

The driver for the passenger
The flights across our skies
The moon that smiles at heights
Professionals for the profession
Everything follows in succession
There's variety of all forms
Yet certain order in these jobs.

LAW 4:

Everything in the universe is constantly working with each other, and at each other to achieve the right purpose and result.

LAW 5:

Everything in the universe has certain amounts of energy usable at a time.

LAW 6:

Every law in nature has other laws embedded in it, and would expose itself to further scrutiny if desired.

LAW 7:
God created the universe, and controls its affairs, man the central processing unit, small but mighty by God's every support.

Mind Dichotomy

The mind perceptibly is divided into two parts and categories: the parts are dichotomized into: 1. (A) Upper/Above (B) Lower / Below.
Its categories are:
2. A. Right side: Upper & Lower
B. Left side: Upper & lower.
3. Box of understanding (4) The deep (5) The canals.

Mind Engineering

In this section, the mind engineering would be in regards to the specimen of mind part, categories A and part B or combinatively categories A and B; functions, faults, corrective measures and spare parts. This is the dichotomy and engineering process of the P.M.Aku's structure of the creative mind.

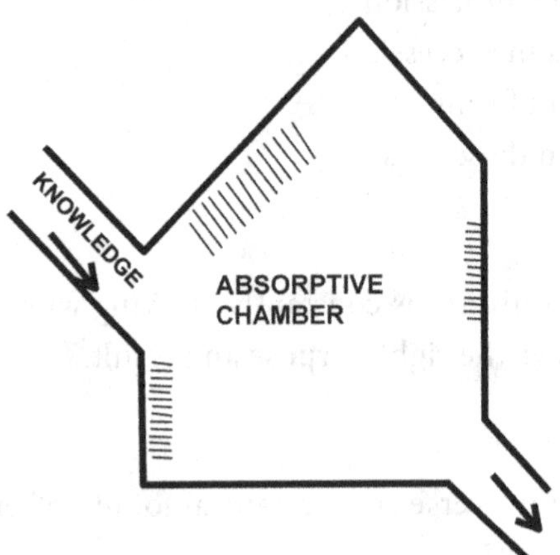

1. Specimen Tittle:
   Absorptive chamber
   Categories A& B:
   Upper /Above; left side upper.

Functions:
1. Absorbs every knowledge allowed by man (its possessor) and sends it to the box of understanding for process in furtherance of use or wastage.

Faults: May absorb wrong knowledge.
Corrective Measures

1. Though no knowledge is a waste, yet unnecessary ones may be discarded.
2. Proper knowledge must be sought from good books, on line resources, experts etc.

Spare Parts
Orientation and upgrading of ones's knowledge through useful practice and personal development resources.

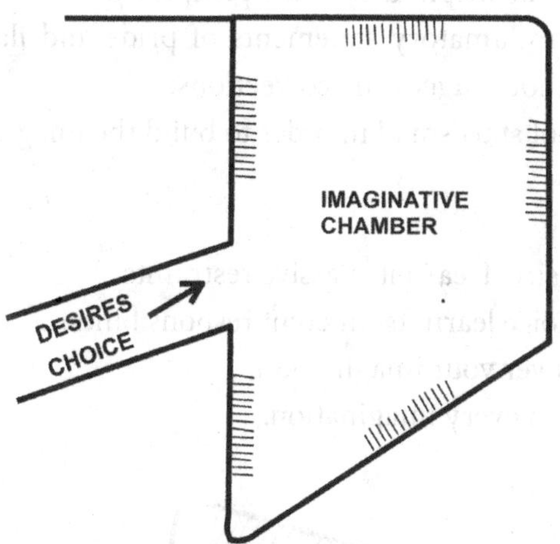

1. Specimen Tittle:
   Imaginative Chamber
2. Categories A&B:
   Lower /Below ; left side lower.

3. Functions:
1. Takes care of the powers of vision, inspiration, formation of mental image and the ability to think.
2. It absorbs every image guided by its acceptance of one's desire and choice made.

3. It is the custodian of creative powers and resourcefulness.

Faults:

1. Dangerous when overtaken by arrogance.
2. Deadly when occupied by evil tendencies such as robbery, murder, destructive instincts.
3. Tends its possessor towards destruction when guided by irrational use of unguided desire for power and fame.
4. Purposeless when dreams occur without corresponding values such as work ethics and endurance through faith and hope.

5. Corrective Measures:

1. Must be guided by godly virtues such as truth concerning trust in God as the absolute maker and ruler.
2. Must learn to be inspired by godly prosperity.
3. Must avoid inflamatory statements of pride and flattery; excessive criticism without suggesting corrections.
4. Dream big but start small in order to build the imagination sensibly.

6. Spare Parts

1. For every desire, L earn its passive restraints.
2. For every choice learn its inherent responsibilities.
3. Self control over your imagination.
4. Fear of God in every imagination.

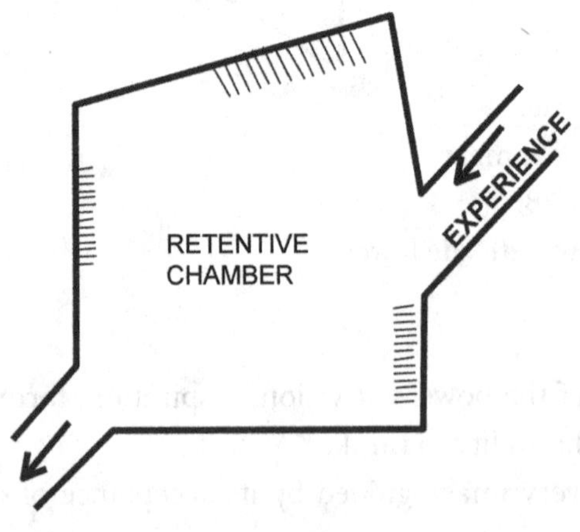

Specimen Title: Retentive chamber
Categories A&B
Upper /Above Rights side upper

Functions

1. Maintains possession of experience in a particular place, condition or position as desired.
2. It is the memory chamber.
3. Sends every experiences to the box of understanding for processing.

Faults

1. May pick and retain negative experiences.
2. May be dulled by poor development and forgetfulness.

Corrective Measures.

1. Learning from negative experiences.
2. Discarding of negative experiences.
3. Mixing with positive minded people.
4. Maintain proper reflections in outlook of life.

Spare Parts:

1. Develop optimism for every pessimism.
2. Always grease your memory by applying the right information.

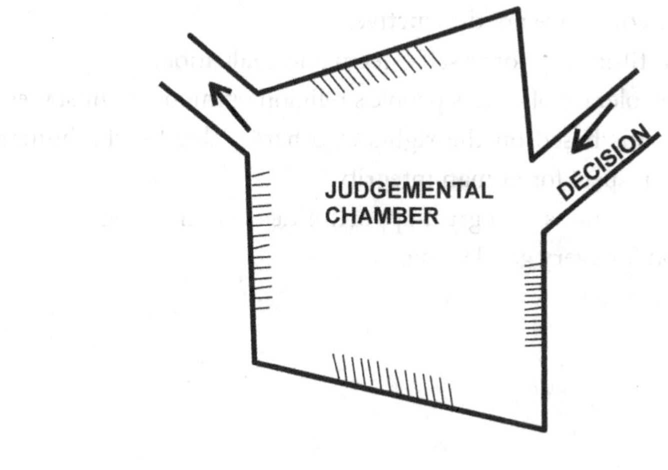

<u>Specimen Title:</u>
Judgmental Chamber.
<u>Categories A & B:</u>
Lower/Below; Right side Lower.

<u>FUNCTIONS:</u>

1. Helps in the mental ability to form an opinion, distinguish relationships, or draw sound conclusions amounting to decisions.
2. Sees to the formation of opinions and estimates after due considerations are processed by the box of understanding.

Faults

1. Subjects to wrong opinions due to narrowness of mental aptitude.
2. Apt to be over critical about issues when not properly informed.
3. Arrogance could make one lose respect for human dignity during any considerations.

Corrective Measures.

1. Fear God as the greatest judge.
2. Judge by acts, upon the prerogative of mercy; and not condemnation
3. Be sensible while forming opinions.
4. Respect yourself and value others opinions and views.

Spare Parts

1. Learn to be corrective not destructive.
2. Learn the attitude of proper self esteem and evaluation.
3. Learn to be tolerant of others people's religion opinion and mistakes.
4. Form opinions based on the rights of others, value for the human person and respect for human integrity.
5. See purity, happiness, and great opportunities in everything.
6. Believe God for every good thing.
7. Study all the time.

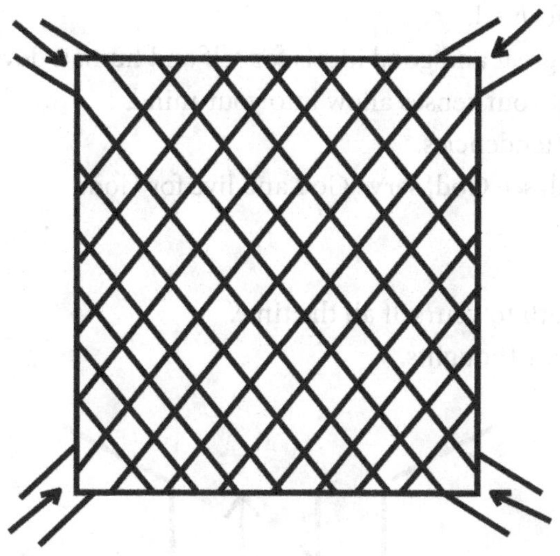

Specimen Title:
Box of understanding
Categories A&B
Central to all part of
Upper/ Above ;Upper / Lower
Rights and Left.

Functions

1. It is the central processing unit.
2. Processes the substances from all chambers, guided by the level of information input, ideas development and resources from the deep and Treston.
3. Controls the final output which results to action.
4. It requires the fuel of information to be creativily active.

Faults:

1. Its works is subject to the principle of garbage in garbage out.
2. What it is given determines and purity for the rights while dirts for dirts.

Corrective Measures.

1. Fear God.

2. Be well informed.
3. Be true to great and good ideals for self and human development.
4. Mind what your senses allow into your mind.
5. Avoid evil tendencies.
6. Think God, see God, serve God and live for God.

Spare Parts:
1. Tell the truth to yourself all the time.
2. Control your thoughts.

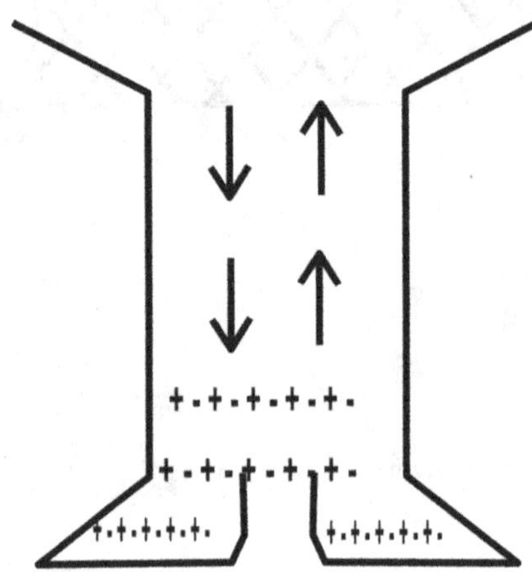

Specimen Title Treston:
Categories A&B
Central, below th e box of understanding; contains deep.

Functions:
1. It contains the ideas travel routes.
2. Contains the creative endowments of a person at bottom and treston -deep.
3. Releases results and answers to in depth questions
4. God's point of direction to man.

Faults:
1. May disturb its possessor if not harnessed of necessary talents mentally or spiritually.

2. May become purposeless if not respected for its resourceful capacity.
3. May become useless if neglected.
4. May become wasted if not developed.

Corrective measures:

1. Must be discovered early through programs and practice.
2. must be developed through education and practice.
3. Must be mindful in using it.

Spare Parts

1. In-depth search from it during problems.
2. Prayers to help you discover yourself.
3. Focus and goal oriented programs must be directed at it.

7. The canals of the information entry and action exit are general service points (G.S.P) to all other aspects of the mind. They work all the time.

SUMMARY

1. The mind is part of human existence which deals with our consciousness.
2. The mind has characteristics applicable to the extent it is allowed or permitted or subjected to.
3. By the application of the Ptolemaic system the mind is the centre of the universe conclusively.
4. The Central Universal process (CUP) is operated by the laws of consistency which are seven in number.
5. The mind is dichotomized into parts and categories collectively.
6. Mind Engineering breaks the various parts of the creative mind with intention to specimen study with regards to specimen title, categories A&B, functions, faults, corrective measures, and spare parts.

With creativity are
potentials harnessed
talents working with time with
patience and perseverance.
We shall one day be remembered.

CHAPTER 9

STATISTICS OF CREATIVE ENDOWMENTS.

Aim/Objectives.

1. To inculcate in the student the logic of Pmakumatics in th determination of creative endowment and potentials in a person.
2. To teach the students to understand persons through the interests of such persons.
3. To be able to interpret a person's nature in the personality predictability perspective (P3).
4. To help the student understand better the concepts of averages applicable to every human person.
5. To Show that attitudes are a product of the mind and present themselves as characteristics or behaviours subject to statistical deductions.

STATISTICS OF CREATIVE ENDOWMENTS

The statistics of creative endowments would in a natural and logical sense be discussed by solving certain Pmakumatical calculations. Based on the principles of understandings which inculcates a provable, but yet logical and acceptability of general behavior ; I shall state it as this:

THE PRINCIPLE OF UNDERSTANDING STATE THAT: CREATIVE Abilities are special endowments placed into a living person: not a morals nor inanimate objects and sustained generally by a certain points of average applicable to every one but depends upon its level of significant development.

Based on this principle we may deduced two points of emphasis to discuss this important statistic.

(1) Every person is endowed to a minimum point of creative Average.
(2) There is the innate tendency to grow upon the level of endowed average (in a person) to the point of recognizable excellence.

Morever, the statistics of creative Endowment (S C E), shall discuss later the various issues of discovery in individuals in relation to how these affect personalities.

(B). CREATIVE AVERAGE

According to the Essentials of statistics in marketing, "it is necessary to stipulate certain properties which an ideal measures of average should possess. Intuitively the following properties seem reasonable;

1. It should be easy to calculate (for practical purposes).
2. It should be capable of objective definition (i.e its value should not depend on any subjective decisions by the individuals calculating it).
3. It should make use of all members of the set of numbers (otherwise information contained in the set is lost).
4. It should be usable mathematically in other statistical calculations. For example it should be possible to combine the average obtained from a number of groups of data to gives the overall average been first combined and then the average of the combined groups calculated,"

The principle of creative Average is best applied when determining the level of endowment, personality predictability perspective (P3), but first; we must discuss the educational technicalities and career categories and allocations as these are vital groups in this study.

The following average or groups shall be considered.

(1). EDUCATIONAL AVERAGE:

The world of education recognizes the principles of abilities testing in a child. This inherently has led to the four cardinal points of testing in a person.

(I). MATHEMATICS: This is to test the calculative ability, intuitively, in a child, or person. This subject, agreeably, involves simple and complex problems and the application of certain logical answers to these; based on laws and formulae derived from principles.

(II). ENGLISH LANGUAGE: Communication is a vital element in our daily living. This subject provides the medium for proper communication and language skills; in order to promote understanding in our social life.

(III). AESTHETICS: The principle, considered most as the essence of existence. It is the study and ethics of social understanding and behavior based on Love,. Interestingly; it enhances the word kindergarten or nursery which by application means "meaningful play under useful supervision". An education without love is non ethical and unacceptable.

(IV). CREATIVITY: This is the study of early detection and possible development of a person or students innate ability to invent by imagination.

The educational average develops the bedrock for the careers averages.

(b). CAREERS AVERAGE.

The global standards recognizes the two distinct points of careers as

(i). Academic - Based: Course of study

(ii). Vocational ; Occupations of practical applications.

Based on these; careers the world over was classified into the following eight categories and sectioned into groups as shown below:

A									CARING OCCUPATIONS
B									SCIENTIFIC WORK
C									JOBS REQUIRING SKILL WITH HANDS
D									JOBS USING ENGLISH
E									CREATIVE WORK
F									FIGURE WORK
G									OPEN AIR/PRACTICAL JOBS
H									OFFICE BASED / CLERICAL WORK
	1	2	3	4	5	6	7	8	

The section- graph indicates the person or individual's interest based on collections from the interest / hobbies classified data base. Please see the student's career analysis program (SCAP) © sample.

This grouping helps us to deduce the personality perspective of a person, as well as calculate the level of endowment in an individual.

Rather than use the conventional mean, median and mode, the creative statistical calculations of the level of endowment (L O E) in a person recognizes the AVERAGE, MEDIUM & HIGH, which we shall discuss later.

(C). GROWTH UPON ENDOWMENTS.

In order to understand the basic principle of growth upon average endowment or creative abilities as stored in a person, let us consider these statements made in the book "Current Biography" 1964, published in the US about two personalities who have earned excellence in their various field of endeavor in different regions.

1. MASSEVITCH ALLA (FEMALE) - RUSSIAN.

Born Oct: 9, 1918.

Career: Soviet Astrophysicist; University Professor.

"The Erudite shepherdess of home- made Moons, is the epithet once used by the Saturday review to describe the soviet space scientist Alla G.Massevitch …

Her research problem concerned the structure and internal energy sources of giant red stars. At that time,..... Soviet scientists up to the end of World War II had Little Experiences in the construction of Stellar Models, and they had almost know automatic computers. Praising her "apparently" inexhaustible energy "and " extra ordinary resilience ". Her success has been "determined inmate intelligence and strength of will"

2. BILL MAULDIN (MALE) - AMERICAN.

Born: Oct. 29, 1921

CAREER: CARTOONIST.

"The most famous cartoonist to come out of the world War II, Bill Mauldin has become one of the America's major pen and ink commentators on the social and political scene....... After a decade of and error in converting in talents to peace time use, Mauldin found his trial new niche in 1958, when he became editorial cartoonist with the St. Louis post- dispatch. Since 1962, he has been doing his editorial cartoons now more Maudlin working credo is that a cartoonist must either have a basic sense of responsibility and dilute it with mischief or be mischievous and temper it with a sense of responsibility ".

The essence of creative endowment are:

1. Productivity; achieved upon its full development.
2. Store of value based upon choice of the holder.
3. Direction - finder to the basic problems of our common existence.

4. Proof of uniqueness of every person.
5. Appreciation of our differences as a basics for unity of purpose and existence.
6. Appreciation of the almighty as the father of variety; the strength of the will and determination to succeed.
7. To live profitably in our endeavours.

TERMS NECESSARY FOR CALCULATING ENDOWMENTS.

1. PEAK OF ASPIRATION:

Every person has within his/her consciousness a certain amount or level of ambition. Ambition is borne out of the need to succeed or excel at one's chosen career, vocation, purpose or calling. When one's creative endowments has been purposely discovered, the tendency to make the most of it successfully begins to grow within him, thereby driving him to an unimaginable height and extent. This is the drive towards the point of perfection, excellence and maximum development. This point is known as the Peak of Aspiration.
The peak of aspiration may be defined as the point of godly perfection inherent in every endowment in a person, with a continuous room for improvement but never attained throughout one's life. It is measured by 1000% IE (infinitum Endowment). The word infinitum means endless. This implies that experimentally no person is perfect but is expected to improve. It also justifies the "Continuous Learning Factor" i.e the compulsory need for continuous learning throughout one's life. The peak of Aspiration (POA) is attributed to the Almighty God who is the father, and creator of all endowments, and its carriers.

The extent of greatness attained by a person during the expression of his in born endowment is determined by him (internal factors) and due less to the economy (external factors). Many books written and unpublished for a present harsh reality on ground (external factor). This applies to the need for profitable relationship system available.

2. LEVEL OF ENDOWMENT:
The level of endowment (L O E) is defined as the natural quantity of creative deposits in a person at the time of creation. This may be deduced through the chosen interests a set of prejudices (likes and dislikes) decided by each person. The SCAP- interests data base decides this factor. Here, 100% is used as an Average Point (AP) while the 1000%IE is used as the POA. This is because man's abilities supercedes the mathematical excellence point of 100% Provably, man has proved his works much more better than the former each time.

CALCULATIONS OF PERCENTAGE ENDOWMENT. IN A PERSON / INDIVIDUALS.

1. These calculations stem form the following criteria that each person's endowment is measurable, in simple terms by deductions of the section/ interest graph.
2. All endowments exist in the present (average) level and has the tendency to rise to its peak in a life time.
3. The point of rising to a limit is called the PEAK OF ASPIRATION, measured by 100%IE (infinitum Endowment).
4. Since all human endowments are limitless, the 100% becomes a bench mark or average of measurement.
5. Scores are assigned to each category in order of importance as following.
   CARING (LOVE) = 20%
   SCIENTIFIC WORK = 10%
   SKILL WITH HANDS - 10%
   JOBS USING ENGLISH - 10%
   CREATIVE WORK = 20 %
   FIGURE WORK = 10%
   OPEN AIR/PRACTICAL JOBS = 10%
   OFFICE BASED/CLERICAL WORK = 10%
   PEAK OF ASPIRATION (POA) = 1,000%

Since " CARING" involves love, perceived as the centrality of all humanity and general relations deemed important in every endeavor: and CREATIVE WORK is the point of endowment stored inside for the expression of this

LOVE or CARE; their allocations both double other categories or groups which are external or outer in nature for the natural is best in the creation of originals. As:

Level of Endowment
= Total Pointing Arrows
  Average point
I.e = L O E = T P A x POA
          AP
Or better still:
L O E = T P A x 1,000% + IE; better expressed as 1,000 %
          100%

Where I represents infinitum or Infinite potentials for growth and development influenced by choice and environment and IE means Infinitum Endowment.

8. "Resourcefulness" is not determined by the level of endowment but by the nature of graph and inclinations of a person.

9. The natural Endowment point is stated in the following categories and determined by the calculation of the level of endowment as:
Average (100% -300% IE); Medium (301% - 600%).
High (601 % - 900%IE).

10. With respect to the L O E; a person whose scores ranges into or above 900% +IE shall be awarded the L O E of or equal to 900% IE + because no person can attain the total point of godly or creative perfection expressed as 1,000% IE, we may only get better.

11. The peak of Aspiration (POA) is central to all parts of intellectual and vocational abilities in a person.

Example 1:
A graph of interests, after plotting has its arrow of highest tendencies on the creative; office based and jobs using English Categories, Calculate the individual's level of Endowment.

Solution
Creative Allocation = 20%
Office Allocation = 10%
Jobs Allocation = 10%

$$L\,O\,E = \frac{T\,P\,A \times 1000\%}{A\,P}$$

$$= \frac{40}{100} \times \frac{1,000}{1}$$

Description MEDIUM:

This calculation is used in career analysis: while the graph determines the personality predictability per- spective (P3).

PERSONALITY PREDICTABILITY PERSPECTIVE (P3)

Basic Principles:

1. The personality predictability perspective (p3) is obtained from the graphical analysis of the interests data base.
2. The interest data base present a culmination of habits, prejudices and behavior expressed in his choice.
3. The graph has " creative WORK atits" center of perspective, but is divided into three parts: (A) LITERARY INCLINED. (B) TECHNICAL INCLINED © LI- TECH.
4. The greatest inclination of interests point to what sort of a person one will most likely be.
5. With or without creative work: the greatest direction of arrow point to the individual general inclinations.
6. The P3 draws its conclusions from the principle that such person could be exhibiting a certain behavior presently or would do so in the interest future; proved from the fact that the chosen individual interest from the interest database represents activities presently appreciated as well as those preferred for future; perhaps when left to independence and the exercise of one's free will.
7. The P3 is subject to environmental factors.

8. Everyone is endowed, special and unique.
9. Deductions from the section- graph describers a person's uniqueness in terms of the levels of expression as measured by the low medium and high relativity to nature of graph; resourcefulness, socials, Brilliance (Dema) creative aptitude (crap); practicality Rating (prat): general Inclination of person (GIP) and description of level,
   The personal attitude chary (PAT).

GRAPHICAL NATURE.
Inherently graphs are classified into four according to number of double heads:

1. SIMPLE: contains four arrow heads on count.
2. HIGHLY SIMPLE: contains between one to three arrow heads.
3. COMPLEX: contains five heads in number.
4. HIGHLY COMPLEX: contains above five arrow heads.

ASSIGNMENT: Define the nature of your personal career Analysis in your own description.

Further Explanations of the P3
The point referred to, and comprising the personality predictability perspective (P3) shall be further discussed perhaps with few historical references.
This would aid the further understanding of human attributes as deductive from the graphical analysis, observation and results from the life experiences.

1. Level of Resourcefulness:
To be resourceful implies the clever and imaginative strength of a person in dealing with difficult situations. This would mean the tapping from a mental store-house of ideas (resources). Resourcefulness also refers to the ability to deal with a situation effectively. Most times, this would be revealed by a complex or highly complex graph of analysis. Highly resourceful people are those who have lots of understanding especially based on experiences from a wide nature and general source.

2. Socials Level:
A comprehensive study of highly creative people often has proved that discretion has often being associated with them. There is always a sense of urgency around them; the need to accomplish this job or that form or nature it takes; as their hobby, and everything. As Georgia O' Keefe once said that her paintings were her children...."Always you are aggravation so that you can get at the painting again because that is the high spot, in a way it is what you do all the other things for. Why it is that way I do not know.. The painting is like a thread that runs through all the reasons for all other things that make one's life." She quoted this in 'Art USA Now when she was trying to explain in different terms what painting means to her as compared with other activities such as seeing friends, shopping, traveling, and gardening. (Reference to current Biography, 1964). From my experience as a teacher, I have come to view pupils and students who are exposed to little social life as prone to higher level of creativity as often their interests in other environmental issues tend to be lower. This is also provable from the fact that children who go through private classes tend to perform better than those who just go through the general classes. This calls for more research.

A complex graph structure would imply a low, medium level of socials. The high level of resourcefulness here, is an indication of an internal struggle of ideas, either known or unknown yet, to this person leading to the process of internal turbulence or mental turbulence within this person. Such a person needs a quiet guidance and monitoring in order to be discovered. Books, news, stories, ideas- talk would quickly be a remedy to this person's troubles. A sign to be watched out for is his inability to talk all the time, a situation I refer to as positive talk trouble. (PTT). This is positive because he would either have too much to say at a time or would not know how to express his ideas at that point in time. This is applicable because his strength lies in critical thinking. He must be stimulated to talk and relate with others.

BRILLIANCE / INTELLIGENCE RATIO: (B IR A).
Brilliance refers to a high level of intelligence with respect to academic work or intellectual application. Intelligence is the capacity to acquire and apply knowledge by the faculty of thought and reason, with regards to the superior

(creative) powers of the mind in facing the various challenges which life may pose in or outside the walls of schooling. This is what makes great ideas highly unconventional. You can agree that the intelligence quotient (IQ...) Age, expressed as a quotient multiplied by 100. While the brilliance level may be deductible from the subject choice chart, and measured by the parallel points of subject enjoyed and subjects good at, the nature of graph would reveal the intelligence level of the person. But in creativity studies, the emphasize is on intelligence level with minor regards to subject score. Nevertheless, theories are still important to learning.

4. Level of Practicality Rating: (PRAT).
Logically and intelligent person is a highly practical person. The more complex a graph the more practical. If two separated to conduct a study; it would be discovered that some are more practical than the others.

5. Level Of Decision Making Ability (DEMA).
People with a high resourcefulness level, interestingly have been noticed to have a low, or medium ability to make decision. This is often the result of having so many internal choices to choose from. They would often require extra time or assistance before a final arrival at a decision. The strength of this ability is the iron endurance that accompanies the very decision made. Since they are not easily convinced at the slightest point, they take every point into consideration, leading to a wider consultation. If not well understood, people would refer to them as slow and inquisitive. But once, their mind is decided they go through with the processes and results.

6. Level Of Creativity Aptitude: (C R A P).
When people are highly resourceful their creativity aptitude level tends to be high. They are not limited by environmental factors, instead they are stimulated by it. This does not mean that people with a low or average creativity aptitude are not creative, rather their is much more to be subjected to further development. This could be done through an extra exposure to principles and other subjective measures like prayers, extra reading and general information sharing. This was the experience of Dr. Maria Goeppert Mayer, a professor of Physics on the San Diego campus of the university of California and a world authority on the structure of the atomic nucleus. She was the second woman in history to receive a Nobel

prize Award for Physics on December 10, 1963, sixty years after Marie curie received hers in radioactivity. She developed her creativity aptitude under the influence of her husband a professor friend. Creativity has no failure rates but can be developed from average level

General Inclination of Person (GIP)

I have discovered and divided creative tendencies into three kinds of inclinations:

(1) Literary: these are people whose creative power tend towards the Arts -Poetry, prose, Play, Theatre, religion, music and all its forms. Exposing these people or students to literary terms would easily be grasped by them. Literary criticisms and legal disputes would put them at their necessary heat of curiosity for logic and rational thinking. I would refer to these set of creative people as literary or Lits.

(2) Technical: these are the people with a mathematical acumen, mechanical mindset, technical understanding. These people work by laws and strict adherence to principles. They grow to become physicists, creative chemists and creative techs.

(3) Li- Tech: this might sound very strange. I have discovered that certain people are multi- talented or multi gifted. They have a knack for the combinative details of the literary philosophy as much as they do have it for the technical calculations I now state here that this is my personality inclination. This is surprise why this work that creativity is not narrow but all encompassing. I had studied the sciences during my secondary school days; perhaps not so brilliant at them ; but I could not help but appreciate the method of deducting formulae and applying laws to workable techniques. I understood them but could not pass the required grades. What mattered to me was the method of practical applications. To me Physics was real, chemistry was practical, Geography was appealing, Biology was jargonized, English was sweet, Economics was about finance control and hunger management, Accounts - a mystery till date since I knew how to calculate my profit and loss and that only scared me but mathematics, I discovered to be interesting.

To prove the issue of Creative multiplicity, I had discovered that I could apply mathematical thinking to writing poetry and solving personal legal tussles. I call these set of people - the Li-techs because it is the combination of the literary and technical abilities creatively.

Historically, let us try to prove this:
Name: Julius Robbert Oppenheimer
Born April 22, 1904 - Physicist; Princeton, New Jersey, USA.

1. During World War II, Oppenheimer directed the Los Alamos Laboratory that built the first atomic bomb and after the war he gave much of his time to advising the government on atomic energy matters.

2. He took an early interest in mathematics and chemistry, and by the age of eleven he had amassed a rock collection and had become the youngest member of the New York Mineralogical Society. Later he had his own twenty-seven foot sloop, which he named Trimethy in honour of trimethylene chloride.

3. Beginning with one graduate student, he gradually built up at Berkeley the largest school of graduate and post doctoral study in theoretical physics in the United States. So devoted were his students that they adopted his mannerisms and accompanied him to his teaching duties at Cal Tech after the term at Berkeley was over so that they might continue to work with him. Men who studied with him made significant contributions to physics and later held important teaching posts throughout the United States.

4. In addition to teaching, Oppenheimer assisted in the management of the Physics department, the selection of courses and the awarding of fellowships. He also carried on his own research.

5. Far from being a narrow specialist, Oppenheimer kept up with other Sciences and with the Arts and Literature. He read classics, novels, poetry and plays; at the age of thirty, he added an eighth language - San-skrit - to his reading repetoire so that he could study Hindu scriptures in the original.

6. During the early 1930's, however, his interests did not extend to contemporary society; he had no telephones, read no newspapers and did

not vote until the elections of 1936. About that time, his social conscience began to develop. Stories of the mistreatment of German Jews (some of whom were his relatives) and the plight of his students who could not get jobs during the depression made him aware of the importance of politics and economics". (As adapted from current Biography, 1964).

When such talents and abilities are found in a child, it must not be discouraged, neglected or abused. These children learn the art of responsibility and economic management earlier, their understanding quite tends to speak more advancely to their hearing. Their intelligence quotient is often high and their imagination is often fired before the normal age of logical reasoning. Sometimes, they are quiet, sometimes they are not. Check out brief ly the story of Andrian Nickolayev; born Sept 5, 1929; a Soviet Cosmonaut.

Nikolayev Andrian is married to Valentina Tereshkova, the world's first spacewoman who was trained by Alla Massevitch (earlier mentioned in this book). He trained with Yuri Gagarin and Gherman Titov... Flying has fascinated Nickolayer since his early childhood. According to one account, he threw his village into an uproar at the age of eight, when after having seen an airfield for the first time, he climbed a tall tree and announced that he was going to fly down. With some effort, villagers persuaded him not to carry out his plan. Nickolayev's Childhood dream became a reality when in 1954, he graduated from an Airforce Pilot School. Once while a Cadet at that School, he crash-landed a jet plane instead of parachuting.

8. Level of Endownment (LOE %)

The calculated level of endowment, as obtained from the interests graph proves the extent of talents inherent in this person. It is measured in percentage with regards to the average point (AP) of 100%; and the Peak of Aspiration (POA) of 1000%. The A.P. proves that man is not a subject to the mechanics of matter. He is the creator of machines, he is not a machine. He has aspiration to achieve as he strives to grow in daily fulfilment of his individual purpose. This peak of aspiration is defined as the highest limit of his aspirations and expressed as the 1000% IE mark.

This mark is signified as the perfection point in man's bid to grow towards his God. Although no one is perfect or prove himself perfect; yet it becomes most advantageous to us if we continue to allow for more sincere learning and tolerance for our fellowmen as well as appreciation of our common existence and collective good. Interests are consisting of choices and prejudices as well as strengths and weaknesses. We must owe ourselves the duty to promote our strengths and covert advantageously the points of our weaknesses.

9. Description of Levels:
Description, here would imply the points of average, medium and high. Here these are expressed not as failure rates or pass grades. It is simply an indication to prove the possible development of such level of endowment. From this description, we may deduce the following.

(I) Level of endowment is subject to further development.
(Ii) Level of endownment is not an expression of failure rates or pass grade.
(Iii) It proves that man has the continuous capacity for growth and development towards Godly fulfilment.
(Iv) No one is perfect, but must grow towards perfection.

10. General Interest Level (GIL)
This is the calculable level of general interest obtained from the personal attitude chart (PAC) or moreso, the psycho-analysis test. By the PAC, the student would be measured against his TTL, energy level (upsurge and downsurged) etc.
It is notable that the GIL is not static but dynamic since choices may change in subject to environmental factors (see laws of adaptability).

Deductions of Formulae for Curiosity.
Curiosity is often accompanied by an internal flow (upsurge or downsurge) of energy. At this point, certain amount of heat is transferred from the brain cells to the mind.

NOTE: Energy is the ability to do work (any work at all), while heat of a body refers to the level, or quantity of the temperature of a body.

By Applicative Principles:
Physical energy is measured by the unit joules, J while creative mind energy (CME) shall be measured by the curiousity joules, (CJ).

Using the Principle of Quantity Analysis:
The instrument used for measuring temperature (the amount of heat physically) is the thermometer. The quantity of Creative heat or curiosity heat is measured applicably by the Thermometric Thinking Level (TTL) measured by the mind, understood by physical attributes or characteristics. This pre-supposses the principle that the mind is the thermometer of all interest and creative processes.

The deductive formulae of Curiosity helps us to understand the following:

1. The psychological operations of the mind with respect to issues such as enthusiasm, depression, expressions, anger, optimism etc.
2. The extent of success or failure of a person's inspiration, aspiration in relation to creativity.
3. The percentage % formular of the TTL.
4. The judgment of a person by his qualities and resultant actions.
5. The development of a potential in a person by the singles and multi-diversification method.
6. The over all level of interest in a person.

However, interest and curiosity play as complementary factors to each other, yet they are distinct in operation. While interest points to a journey, curiosity takes the person through the journey with defined paths and stages of such journey.

Interest level and its calculative application is also a big factor in understanding an adult. Next we shall study these factors and see the extent of interest an adult mind is also likely to develop in a purpose. This calculation is used for adults only because at this stage of life, the mind is developed to the stage of decision making. Unlike the limited scope at junior stage, it is a determinant factor for a lifetime interest in issues bothering on his/her vocation. At adult level, the psycho analysis test or personal attitude chart is conducted compulsorily. It would help employers understand what goals to set, what role to fill in for a prospective employee, as well as foster an ideal appreciation of statistical records for the dynamic nature of a people, say over a period of three years.

By these the general SCAP is necessary for educational purposes such as Class placement, student's behavioural attributes (p3), Universities and faculty placements, Senior Secondary School Class placements, Security trace and violations of natural ability (misplaced priority) etc. While the PAC is conducted at the Final year of a higher Education Training for employers and personal records compulsorily, as part of the Curriculum Vitae (CV).

Calculating the level of Interest as a person has in a thing is a determinant factor on the extent he / she likely to succeed in a thing or purpose.

In order to calculate this, two major factors are considerable:

(I) Calibrated Curiosity Levels of interest table (The Pmakumeter Guage)

(Ii) The quartet (1/4) principle

1. The calibrated Curiosity Levels Table (CCLT) / C2 LT.

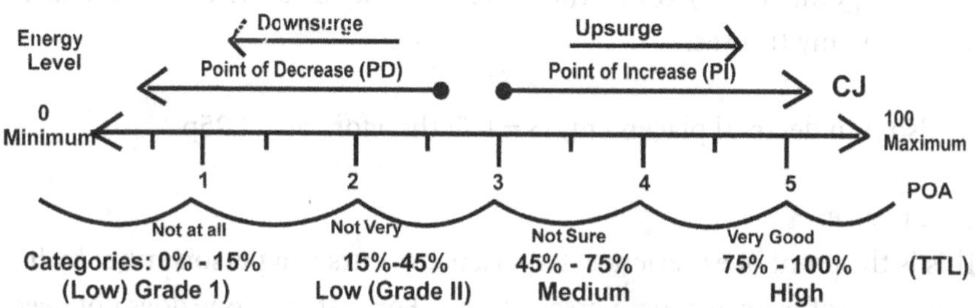

2. The Quartet Principle:

The quartet Principle is guided by the calibrated (curiosity) levels table (C2LT)

The quartet principle states that the combustinative factors of experience at a time and place determines the level of one's interest and is a determinant in the success or failure in such purpose.

$$\text{Interest} = \frac{\text{CU x TM x PL}}{\text{OC}}$$

But using the Quartet Principle:
Curiosity is obtained by using the points of a stated level from the C2LT obtained from the personal attitude chart (PAC) of adult - SCAP.
Time is considered ¼ x 24 hours since no one expends considerably, more than 6 hours a day doing just one thing.
In a day, man works, eats, drinks, sleeps, wakes, talks, baths, drives etc. This defines the fact that he will not subject the totality of a day to just thinking to the total neglect of other aspects of living.

But 6 hrs x 60 minutes x 60 seconds
= 360 x 60 = 21,600 seconds average
Or expressed as 21, 600 SAV

Place & Environment
The earth is also guided by the quartet principle:
The four cardinal points being the North, South, West and East. This implies that the environment one lives or exists at the time of interest is only a quarter in the cardinal points. That is, no one ever stays all over the cardinal points at any time. Therefore, place defines the environment of interest at any time as:

¼, also if in decimal placement; ½ = 0.5; therefore ¼ = 1.25p

Occurrence:
This is the point of experience. Experience occurs to a person either by his own terms of from another persons terms. For Instance, one does not need to stand before a moving vehicle to have the experience of accidental death. He is expected to learn from previous casualties or records. Knowledge is a form of experience. Experience itself is knowledge. The quantity of experience in one is considered as ¼ litres of one's learning. Therefore, the quartet principle is also applicable here as: Occurrence = ¼ L

But experience is a product of a person and another person's. No experience is total or else, no need exists for learning.

Applying the interest formulae using the quartet principle is thus:

$$I = \frac{C \times 21{,}600 \times 0.25 \times 1}{4}$$

But curiosity Cu is obtained from the C2LT.

Practical Application

1. The interest formula is used to determine the depth or extent of a person's interest and eventual success or failure in a psychoanalysis or personal attitude chart.
2. It shows the extent a person knows and understands himself.
3. It is used only in the Adult Category of the Student's Career Analysis Program.
4. It is only effective at Adult stage after being studied.
5. After the highest point of Cumulative average is obtained from the psycho-analysis chart, this is measured against the CCLT in order to obtain the curiosity of such one, and as well its category.

Class Exercise:

A woman of 26 years was discovered to have a total scores of 32, 27, 15 in the Personal Attitude Chart. Determine her total level of interest.

Solution: Total Section Scores (TSS) - Single Higest Score (SHS)

TSS - SHS = CU --- (i)

32 + 27 + 15 - 32 = CU --- (ii)

74 - 32 = 42 Cu --- (iii)

2. $\frac{Cu \times 21,\ 600S \times 0.25 \times l}{4}$

$$\text{Interest} = \frac{42 \times 21{,}600 \times 0.25 \times 1}{4}$$

$$= \frac{226{,}800}{4} = 56,\ 700 \text{ or } 57.0 = 57\%$$

When measured on the C2LT / TTL categories
She has a medium TTL

I.e. TTL = Medium

Example 2:
If a Civil Servant, male aged 37 Years had the scores of 30, 18 and 24 in his PAC
Determine his level of interest.
Solution:
TSS - SHS = Cu
30 + 18 +24 - 30 = Cu
72 - 30 = 42Cu;

$$I = \frac{42 \times 21{,}600 \times 0.25 \times 1}{4}$$

= 56,700 = 57.0

TTL Category = Medium.

Example 3:
Calculate the level of curiosity for a person aged 60, whose interest level is 84%; and
His PAC scores are a total of 65.

$$I = \frac{CTP}{OC} =$$

$$C = \frac{TP}{IOC} = C=$$

$$C = \frac{21{,}600 \times 0.25 \times 1}{84 \times 4}$$

C = 20.25 x 65 = 16.25
= 16 Cu.

SUMMARY

1. The statistics of creative endowment is guided generally by the principle of understanding.
2. The principle of understanding states that creative abilities are special endowments placed into a living person; not animals or inanimate objects, and sustained generally by a certain point of average, applicable to everyone but depends upon its level of significant development.
3. The statistics of creative endowments is guided by the principle of creative average, educational average, careers average, growth upon endowments and the calculation of creative endowments in a person.
4. Level of Endowment (L O E) is calculated thus:

   $$L\,O\,E = \frac{\text{Total Pointing Arrows x Peak of Aspiration}}{\text{Average Point}}$$

   $$\text{I.e } L\,O\,E = \frac{T\,P\,A \times POA}{AP}$$

Where AP = 100%, PO A = 1,000%IE.

5. The P3, like the L O E is deducted from a creative study of a person interest graph.
6. The P3, interprets a person's or likely behavior in respect to factors such as levels of resourcefulness, Bira, Prat, Socials, G I P, L O E, and Description, Nature of graph, creativity Aptitude (crap) and Dema.
7. The deductions of the formula for curiosity is a factor necessary to calculate interest.
8. The basic calculation of curiosity and thereby Interest helps in the interpretation of the P3; thereby eliminating the employee interview process.
9. The Pmakumatical instrument (Cc L T)is used to energy level (upsurge and down surge) and is practical.
10. This aspect of calculations deals with the internal creative content of man; white general mathematics studies and calculates physical quantities.

CHAPTER 10

THE MASTERS

Aims / Objectives:

1. To list the names of certain persons who have excelled in creativity.
2. To encourage the student to conduct personal research into the various works of these people.
3. To create confidence in the student to grow himself in the opportunities which abound in creativity.
4. To prove by this listing that creative endowments have made people great, regardless of gender or skin colour.
5. To confer through this page the revered title of sage to those people whose Creative works are considered outstanding.

"The purpose of education is to fit us for life in a civilized community and it seems to follow from the subjects we study that the two most important things in civilized life art and science".

In this really true? If we take an average day in the life of the average man we seem to see very little evidence of concern with the sciences and the arts. The average man gets up, goes to work, eats his meals, reads the newspapers, listens to the radio, goes to the cinema, goes to bed, sleeps, wakes up and starts all over again. Unless we happen to be professional scientists, laboratory experiments and formulae have ceased to have any meaning for most of us, unless we happen to be poets or painters or musicians or teachers of literature, painting and music the arts seem to us to be only the concern of school children. And yet people have said, and people still say, that the great glories of our civilization are the scientists and artists.... Any way, both the artist and the scientist are seeking something which they think is real. Their methods are different" (John Burgess Wilson).

The brief passage bears testimony to man's quest for inner satisfaction. Today, in this book, I say that creativity is the answer. Not even the worship of God would be rich without a heart working in its line of fulfillment.

This page or chapter, is dedicated to the remembrance of the masters whose toils have made life easier and sweeter, more convenient to live and then to the living, I want to assign honour. It is not possible to state everyone's name, nor is it possible to remember; yet the heroes of creativity shall be regarded as the masters of skills and sages for life.

I am a Nigerian and I bear testimonies, in my time to the persons of Wole Soyinka, Chinua Achebe, Chukwuemeka Ike, Obi Egbuna, Onyeka Onwenu, Sir Warrior, Victor Uwaifo, Oliver De Coque, Bongo Segwe, Art-Alade, Fatai Rolling Dollar, Sunny Ade as testaments to the bearers of creative lights in Nigeria. Welcome to the page of the masters.

From my iconidol and sage William Shakespeare, I honour the likes of Euclid, Pythagoras, Homer, Sophocles, Nehru, Eisenhower, Winston Churchill, Albert Einstein, Madam Curie, Bernard Shaw, George Orwen, Louis Carroll, Edward Jenner, Louis Pasteur, Orville and Wilbur Wright, Langley; Beethoven, Joseph Conrad, Demetrius Kapetanakis, Ernest Hemingway, Lin Yutang, Charles Dickens, Geoffrey, Chaucer, Sir Thomas More, Sir Francis Bacon, William Beckford, T. S. Elliot, Wordsworth, Alfred, Robert Henryson, William Dunbar, Gavin Douglass, Caxton, John Wycliffe, William Tydale, Miles Coverdale, Seneca, John Heywood, Thomas Norton, Thomas Sackville, Kyd, Nicholas Udall, William Stevenson, John Lyly, George Peele, John Milton, Christopher Marlowe, Ben Johnson, Fletcher, Thomas Middleton, George Chapman, John Marston, Thomas Dekker, Thomas Heywood, Beaumont, John Webster, John Ford, H. G. Wells, Sir Thomas North: all these do prove that creativity ought to be a sector; a thriving industry just Banking aviation defense, if we can harness it's potentials. These like and more of them sought for value in life and got it through their own potentials.

Perhaps, I should give you more of these masters: Robert Burton, John Foxe, Richard Hooker, Tolstoy, Scott, Nashe, Deloney Thomas, Edmund Spenser, John Donne, John Dryden, Sir Thomas Wyatt, Sir Phillip Sydney, Samuel

Daniel, Michael, Drayton, Horace, Robert Herrick, Marvell, Thomas Andrew Carew, John Cleveland, George Herbert,.W.E.B. Dubois Richard Craw shaw etc. and the legend himself Michael Jackson, to all these and those of our days, I place upon my honour as sages, though dead, they Jet live; and for the living, they are honored; for it is sweeter in our days to give honour to whom honour is due while alive, rather than singing fabulous poems of eulogy at their grave side. He who achieves does so while alive, the dead may be present, but not enough to show appreciation for good speeches and awards.

The creator brings true changes
the manipulator seeks false praise
the pirate is a thief of few days
greatness to them seems so strange

CHAPTER 11

CREATIVITY & MATERIALISM

Aims/ Objectives:

1. To discuss the school of creativity and the school of materialism.
2. To establish the relationship between creativity and materialism.
3. To show the facts of contemptible narrow-mindedness behind materialism.

Creativity & Materialism

The word materialism strikes the mind of the average person as simply excessive regard for worldly concerns. However, for the purpose of this study, the original meaning of this word is to be regarded with great concern for creativity. Here, materialism is the theory that physical matter is the only reality and that everything can be explained in terms of matter and physical phenomena.

Our respect here stands in our regards and concern to find the relationships appropriate to distinguishing between creative endowments and the existing physical phenomena or matter, mindless of whatever category it may belong. Here, we hope to start the argument between the school of creativity and the school of materialism. Let us first consider the following pertinent questions:

1. Before Henry ford; was there a car?
2. Did the dynano exist before Thomas Edison?
3. Who first preached the doctrine of transcendentalism before Ralph Waldo Emerson?
4. How many machines existed before Eli Whitney's cotton gin?

5. Did Isaac Newton read the principle of gravity from a book; or perhaps the 1st, 2nd & 3rd Laws of motion?
6. Who had invented the Pmakumeter Guage before Peter Matthews Akukalia; or perhaps fathered the complete field of creative sciences?

Two likely answers, I perceive, suit the majority. Firstly, some would quickly and zealously, answer God is the maker of these. Secondly, the others would mention the very names as answers to the questions. If you answered God, then I would accuse you of contemptible narrow mindedness because it is established by facts that;

(i) God made man;
(ii) God authorized man to create by his endowments and intelligence.
(iii) God made the earth and endowed it with the necessary materials man needed to create. Therefore, it is not wrong to say that this person should be named after his works. While man takes credit, then God deserves His glory. Below is the table of relationships between creativity and materialism:

Relationships between Creativity & Materialism

Creativity	**Materialism**
1. Endowments are within the person.	Endowments are in the earth.
2. Brings structures to reality.	Makes structures into things.
3. Controls man by purpose.	Is controlled by man's purpose.
4. Uses materials to create.	Useless until discovered and used by the mind.
5. Creates strategy for sensible use of energy.	Subjects one to hard work.
6. Has no limits.	Is limited by environmental factors.
7. A couple of talents that worth more.	A couple of ideas of determined worth.

SUMMARY

1. People by their creative endowments create and still create things from ideas than from materials.
2. Several relationships exist between creativity and materialism proving creativity more superior than materialism.

CHAPTER 12

WHO IS A WRITER & CREATOR?

Aims/Objectives;

1. To teach the differences between a writer and a creator.
2. To help the student understand the types of creativity present.
3. To prove the various qualities of a writer as compared to a creator.

WHO IS A WRITER & CREATOR?

Who is a writer? Who is a creator? Is a writer also a creator, or perhaps can we describe a creator as a writer? Very well. I shall, in my broad opinion, since it is a recognizable part of my rights of expression, though one expected to be at least reasonable and logical to an extent of simple understanding, for a good lecture should be broken down into simple and applicable principles; try as much as possible to explain this point.

Firstly, let me try to answer with a question or two. Can the creator, for instance, the music composer, or singer, or artist (One who draws); or fashion designer, or a sowing machine designer and operator, carry out any part of his vocation with out making certain notes? If your answer is no, could he reasonably be referred to as a creator? And perhaps, if your answer is yes; then does the making of notes ever qualify such one to be called or honored as a writer? Both of these are distinct and still are the same. Deferent yet still the same. Clearly, a writer is a creator whose life is measured by the values he adds to beauty by dissecting words. He feeds from words sensible words. These words express ideas. Ideas are expressed in words. Ideas of creativity remain abstract, sitting comfortably in the treston of the author, until it is expressed in words, for instance, the word 'Car' brings to mind the idea of a mobility machine, yet of what sense does it make when it has no name expressed in a group of letters to form a sensible word. My research results prove that the writer is not only a creator of books, he is also a creator of words for instance, did you know that the words, "Assassination", and "Bump" were created and used meaningfully, by that sage of the stage, William Shakespeare, whose literary prowess and certain impatience in writing led to a natural creation and immediate use of these words? A careful study of this sage, without excluding others, but a special focus for a case study proved that his wit occurred naturally and spontaneously. Well, what do you say about the complexity of the human brain and its powerful capacity to process information and get the needed direction in solving problems. When primitive problems are solved, the results lead to civilization. What about the creator? Creativity encompasses everything creatable. The creator is a living embodiment of various aspects of creative technicalities, except his focus is dominant upon one side. I shall, no doubt, point that the creator is an inventor, a developer, a facilitator and a maker. Creativity in my sphere,something like my globe, portrays itself in three ways or types.

Types of creativity

1. Dominant 2. Recessive 3. Diversified

1. Dominant creativity: shows itself noticeably in the life of its possessor. Here, greater emphasis referred to as Trained focus (T.F.); is laid upon a particular aspect of his creative potentials. It is also referred to as kinetic or Visible creativity.

2. Recessive Creativity
   Special creative ability which is there, but never used. It may also be referred to as Hidden or potential creativity Here, it's clear that the possessor's interest level is low, perhaps below 15% (See TTL) in chapter 1; for instance, a person to become a music composer due to his interest in musical lyrics yet he does not wish to become a musician or take the credit for singing it. So what options are available to him? It is either he decides to sell it for sums; or he keeps it in his drawer in order to draw a personal satisfaction from its presence.

3. Diversified Creativity.
   I have often heard people make the statement, "Jack of all trades, yet master of none", often said in order to promote a special ability in a single object rather than knowing so much and profiting from none. In as much as this seems highly correct, I stand to prove that the creative mind can do and perform different tasks and know each one so well as to profit him.

   This is what I shall refer to as 'Diversified creativity'. What do you say to yourself, or perhaps, think in your mind when a person is a telephonist, a phone repairer and distributor? The word 'Expert' or 'competence' occurs. This creative mind has no limits. Let us examine the life of Theodore Roosevelt, whose life was noted to have been a president on a wheel chair in the prestigious White House of America. He was, in the words of Lekan Fasina, "a Nobel prize winner, a paleontologist, a terxidermist, an ornithologist, a naturalist, a conservationist, a big game hunter, an editor, an orator, a critic, a ranch-man, a country squire, a civil service reformer, a socialite, a patron of the arts, a colonel of the Calvary, a former Governor of New York, and a U.S. President.' It is credited to this highly creative man the statement that "all we have to fear is fear itself". Now I shall add that the fear to express your creative abilities is the fear to live courageously and die cowardly. Perhaps, our institutions of learning should create a special "faculty of creativity" and offer its recommended modules of this subject book to the students and scholars alike. This is a candid recommendation, in my own opinion, meant for urgent and due considerations. I believe in diversification of creative talents because it trains the mind for distinction, development of mental alertness, the appreciation of variety interests, judgemental ability, sequential follow up of interests, expertise, proper time management and total expressiveness.

Alternatively; a writer is a single-headed arrow while we can say that the creator is headed: one who is the "Jack of all trades; master of all", beyond every comparison, and thriving above the pettiness of singleness, he is highly knowledgeable, practical and total in his view of issues whether considered of great importance or seen from the eyes of triviality. I choose, therefore, this day, to be called a writer, if you like; but preferably a creator; a facilitator.

Obi B. Egbuna, the one whom I describe as the silent shaker, for his words ignite a quiet heat which bursts into flames to accomplish its goal, not minding your stand on the issue; once described the writer in the following statements:

"A writer's life is Hell; A writer is a one-man band of angels. A one-man crew of Lucifers. A one-man family. A one-man university. A one-man compost heap. A one man traffic Junction. A one-man jazz of ecstasy. A one-man choir of blues. A one-man. Everybody's nobody. Writing is the loneliest occupation in the world". I shall say that this description is direct to the writer and indirect to the creator. Let me also state that in my honest belief, one who writes beyond the shores of just one type of literature qualifies to be known as a creator.

Perhaps, we shall say that a novelist, or play Wright, or poet is a writer; but when a person successfully carves a riche in poetry and prose, or the three aspect successfully, she should be called and honoured as, not a writer, but a creator. You see, it makes a more respectable sense to accord elevated platforms to special workers and outstanding achievement; in order for such ones in case, they are short or stout in height; they could see more clearly when the ovation comes; and if tall already, than the public would see them better.

Then, I shall write, this day, how I perceive the writer and now the creator whose personality represents God's own presence, except he fails by the law of pride and pettiness. In my sincere perception, I see the creative person or creator as a theater of doctors, the energy of culture, the arrow-pointer of society, the guide of values, a church of prophets, a street of shining lights, the speedometer of life's vehicles, a waive of news casters, a library of books, the heir to immortals, a raging storm, a calm breeze, a galaxy of stars, a heart of lovers, a forest of trees, a kingdom of princes, the breasts of mothers, a father of nations, a mother of children, a school of teachers, a farm of husbands and a garden of wives; a creator is a war front and a peace-hut; a great thinker, he is always dangerous at the sight of simple instruments- the pen or pencil, the brush or paper, colours of an evening and a nation of strategists.

Intellectuality and civilization owe its light to the labours of the creator. He has the midstouch, the Aku's miracle fingers; an oracle and a definite master of his art, he is also brand in itself and role model, a standard in himself as well as a banquet to eat and be nourished from.

All these described above is what I shall sum up to be referred to as the P.M. AKU's LOGIC OF THE CREATOR OR P.M.AKU'S L.C. This logic stands on the P.M. AKU'S principle of creativity which states that "The creator is a centre of attraction and centers all his requirements in himself".

SUMMARY

1. A writer is a creator but a creator may not necessarily be a writer, since many aspects of creativity exist.
2. Creativity is all encompassing
3. Types of creativity are a) Dominant b) Recessive c) Diversified
4. Dominant creativity is that trait which is most noticeable in a person
5. Recessive creativity is the present but hidden trait in a person.
6. Diversified creativity is evident when a person exhibits may traits, grapples and wanages them well.
7. The P.M. Aku's Logic of the creator states that the creator is a centre of attraction and centers all his requirements in himself.

I am on fire, I'm inspired
is big risk;
I make the world better by my trial
to help men think.

To strive and conquer
is a big task;
I owe the world nothing but
to climb the intellectual ladder.

Imagination; the blesser and punisher
Speaks life is sweeter

Correction; the holder of the docile
Says life can be better
Revolution; of the attitude and mind
Brings to men better heights;

And inspiration binds them all
In strong ties.

CHAPTER 13

INSPIRATION & IMAGINATION

Aims /Objectives

1. To expose the process of inspiration and the strength of imagination in creative process.
2. To prove that inspiration and imagination involve certain processes and activities.
3. To help the student realize the need to build a purpose in creative development.
4. To show the true meaning of vision.

INSPIRATION & IMAGINATION

I shall in a most informal manner begin this chapter. I hope to discuss with you a very simple and yet interesting word which caught my interest during the voyage of the course of my research on the issues which bothered me most, especially on creativity and its relationship to the pre-historic times, and then more so, how other nations became suddenly exposed to certain happenings around them. I bumped upon a word. "Renaissance" In as much as it sounded pleasant to my ears, more civilized in pronunciation and seeming intellectual in its very nature, its meaning also added flavour, that which I desired to know in my work and research. Funny enough the truth behind the title page of this book was that this word first inspired me as a title. But the God of creativity wouldn't let me use it; I hope my frankness concerning my success story doesn't bore you.

According to the American Heritage Dictionary, third Edition, one of my literary oracles; the word "Renaissance", goes thus; 'i.A rebirth or revival. 2a. The humanistic revival of classical art, architecture, literature, and learning in Europe. B. The period of this revival, roughly the 14th through the 16th century. 3A. revival of intellectual or artistic achievement." Another dictionary renders it as "1. The period in Europe during the 14th, 15th and 16th centuries when people became interested in the ideas and cultures of ancient Greece and Rome and used these influences in their own art; literature etc; i.e Renaissance Art (ii) A situation where there is new interest in a particular subject; form of art etc; after a period when it was not very popular "By a careful study, I decided further to probe the word revival and my oracle the American Heritage dictionary renders it as, 'To return or bring back to life or consciousness, to impart neir health, vigor, or spirit to; to restore to use, currency, activity, or notice; to present (e.g., an old play) again".

In my high state of interest, I discovered that someone must have taken the lead; Someone must have bore it in mind to cause a revival; someone must have experienced a sleepless night due to peacelessness which would often come with a new vision, so, where did peace come from? Where are visions born? Why would a person feel so obliged to destiny to attempt the heinous task of trying to make a change offering himself the loneliness and uneasiness of a reclusive life? Their interest offered them knowledge which they sought until they became suddenly aware of the vast opportunities, which lay around them all the while. This state of sudden awareness is what is termed inspiration.

My oracle defines inspiration as, "a. stimulation of the mind or emotions to a high level of feeling or activity, 1. The condition of being so stimulated; 2. One that inspires. 3. Something that is inspired. "To understand this further, the word, inspire; means, To fill with nobble or reverent emotion; exalt. 2. To stimulate creativity or action. 4. To elicit or create in another. 4. To inhale" So, our conclusion is arrived when we say that true works of creativity are often a product of inspiration; pehaps made from a single statement of fact; or a past knowledge duly discovered. That is what I shall call my first Law of inspiration P.M. Aku's 1st Law of inspiration states: "(Creativity is the product of inspiration and knowledge" Since the extent of knowledge is a determinant factor of creative success. Pmakumaticlly, I shall express it as;

Inspiration X knowledge = Creativity or, Is X Kn = C.

When inspiration arrives, comes a mental picture. The mental picture is dependent on zeal or enthusiasm; the belief that it is possible. At this point, the possessor of such mental picture tends to influence the extent of his possibilities by the size of the mental picture in his mind's eye. The mind is a comer, man determines the size and quality of the picture. The mental picture which is formed in the mind of the possessor is called imagination. Inspiration is the connecting like between the intellect (the thinking faculty) and the spiritual source or the muse, which forms the mental picture. Inspiration fires the imagination to produce into the mind; while imagination boosts the intellect and thereby moves the body to action. It is highly interesting to note that since the creator could be influenced to carry out certain activities, perhaps, history may not forgive me easily if I fail to point out that such powers of the muse could be of positive influence or negative. Let's break them down.

NEGATIVE INFLUENCE MUSES (NIM).

These muses or powers tend to influence the creator who is a person to carry out evil or dubious sections or works even though they know their onions. Such Acts as genius in criminality, terrorism, internet fraud, and stunch un-forgiveness, here, are caused by these. The greatest aim usually of negative influence Muses is the promotion of violence and destruction. The creation of missiles and weapons of mass destruction have never promoted true peace. Man is a being of influence, not a trifle of spiritistic caricatures.

The creator must watch the direction of his inspiration voyage as well as concentrate in order to land not only safely but happily too, after his flight.

POSITIVE INFLUENCE MUSES (PIM)

The first mission given to a creator who possesses or is guided by a positive influence muse (PIM) is to promote peace and human dignity, why so? This is done both in his best interest and that of others, A world of such creators of positivism would be a world of true beauty and positive values. The results of such works would be proper administration, resourcefulness, diligence, reward truly deserved and peaceful co-existence, trust, self-

reliance, observance of basic ethics and safety rules, as well as national progress. These are the ones who are truly honoured as sages.
While inspiration is the light of creative vision, imagination is the powrer and the commission bestowed upon the creator to do, act or bring to existence. Next, I shall try to discuss. Imagination as much as possible.

THE SCIENCE OF IMAGINATION

Is imagination best described as a science? Yes, to the best of my knowledge because it involves a process, though for study and involves certain understandable yet obvious principles. Imagination is the science of producing and keeping the mental picture in the mind for a measurable time.

P.M.Aku's Law on Imagination States:
Imagination is dependent on the strength of the desire and the projected image by the mind of the creator.

Pragmatically, this would be:
P.M. Aku's law of imagination

Size of image + inspiration X1,,000%IE
Desire =Imagination (IM)
=S.I+I x 1,000% IE =IM;:. SII x1,000% =IM.
D D

; Where 1,000% is used because it's S.I.+I1,000%IE=IM.
Or SII x 1,000% IE=IM.
Based on my personal experiences, I deduced the following types of imagination or imaginative power as:

Keen Imagination: Also known as normal imagination. It lets your mind enjoy the process of normal peace. Often slow in its movement or progressive pattern, it is the first time a mental picture and its idea occurs in your mind's eye. At this point, it seems poorly feasible nor realistic, but your hold on it keeps it going. Here, fragments of pictures and thought processes occur to the creator

Warm imagination: Also referred to as fast or signal imagination. It keeps the creator mentally alert, optimistic about the idea and helps you develop the zeal enough to begin research. Here, your power of expression about this idea is noticeable. Confidence is gradually built and runs itself into the point of zeal or heating passion.

Hot or Restless Imagination:
At this juncture, the possessor burns internally with fierce passion, clearer understanding of the issue to be discussed or created, Such one tends to speak less and work more. His inner eyes see visions of success as highly realistic; the mind is restless and there your fingers become miracle fingers-ready to work and powerful to bring to reality; as sparks of fire. The remedy for this would be to push out the hot object in your mind to your paper. I shall also warn that here, one sometimes tends to hear voices as though guided by a person or special powers such as higher spirits. The possessor at this level of imagination is unstoppable, unpredictable and not even the noise of the environment, nor difficulties of working under pressure can defeat him. At the end of the work, you are nothing short of excited. Then the heat slowly cools off and then out and will probably surface again only when that work is being remembered, spoken about or referred to.

PURPOSE AND VISION

My oracle defines purpose as "1. An aim or goal 2.A result of effect that is intended or desired, intention. 3. Determination; resolution: to intend or resolve to perform or accomplish" At the point of imagination a decision to achieve a set goal is determined. This set goal is defined as purpose. The goal, here, I shall refer to as the product. Purpose is usually a direction-pointer and helps to build the mind to think better. In my opinion, based on my personal discovery, I shall define three kinds of purpose as:

1. Initial purpose 2. Defined purpose 3. Established purpose.

1. Initial purpose:

This is the first time a purpose or set goal begins to surface in the mind. It is also known as Half clear or Unsure purpose. Fragments of the product or set objective come to mind but will need more research in order to obtain better information about it.

2. Defined Purpose:
This is the next point of purpose where thoughts are clearer, for instance, write a story book with this title or that one, for the primary school children. It is easier to work in this stage because you are moving in the line of your thoughts towards your purpose.

3. Established Purpose;
This is the point of finality when the work is created. It is now readily for use. it has come into existence just as much as you imagined it. Your mind's eye saw it; and your hands created it. Only he can tell how exceedingly demanding it is a task to create. The established purpose brings happiness to the creator because his satisfaction is an inner feeling of contentment, One derived from the fact that he achieved it just as his mind thought it. As a creator, Imagination helps me create a play of reasonable volume within four days, firstly, it used to be two weeks until I improved on that record. Surprises of this sweet nature, though pleasant, but do come with its own headaches, sweat and self denial.

P.M.Aku's Law of purpose states that an established purpose is the result of the sum of the Initial purpose and defined purpose, and without these, has no existence.
Pmakumatically;
P.M.Aku's Law of Purpose
I.P.+D.P. E.P; i.e.
Imagination

I.P. + D.P. E.P.
IM

Since purpose is dependent upon imagination, and relative to the person's intention, we may deduce the genres of purpose as;
A. Real Purpose B. False or Fake Purpose.

A. **REAL PURPOSE:**

Often based on genuine intentions of self development, human dignity and achieving positive results. It is developed upon knowledge or experiences about something, of which opportunities for change are seen, perceived and felt by the purposer.

The Purposer, I define, is one who holds a purpose. This purpose becomes real when it is workable, with a human-development capacity, measurable, teachable, specific and realistic; in terms of paying for its rewards. In creativity, the product is not sold but it's advantages; its process is not payed for but its satisfaction. The knowledge which defines purpose may be complete or incomplete and so research is endless.

B. **FALSE/FAKE/DECEPTIVE PURPOSE:**

This is a purpose found on greedy tendencies such as fraud, stealing, robbery. It's bedrock is illusions. The purposer here, does want to excel but hates work; prefers to be praised but hates to think of the process; or perhaps, ever hates to get involved. The purposer with a false purpose is a 'fake person and soon people will discover his poor attitudes and begin to avoid him. He is no good because he is not reliable.

FACTORS OF PURPOSE

These factors stated relate basically to real purpose. They are objective in scope.

1. Purpose does not die but may sleep temporarily
2. Purpose sees opportunities where others claim blindness or shortsightedness.
3. Purpose respects humanity
4. purpose is a direction finder
5. purpose is a confidence booster
6. purpose builds and elevates the mind.
7. Purpose is a reality
8. purpose solves problems.

Here, I shall define problems in three groups as:

A. Short term e.g. hunger, clothing, shelter, etc.

B. Medium term e.g debts, suitable finance, Simple necessities (e.g. a car etc).
C. Long term e.g Financial security, permanent and personal accommodation, wealth, fame, economic stability and immortality.

ENEMIES OF PURPOSE

1. Yourself: your perception and attitudes to your purpose.
2. Distractions: Unnecessary pleasures and procrastinations
3. Negativity & Doubts: Poor belief, lack of self confidence and little or zeal.
4. Cynical Remarks: complains, negative confessions and personal distrust.

VISION:

My personal oracle defines the word vision as……unusual foresight; a mental image produced by the imagination;

Inspiration breeds imagination. Imagination breeds purpose and vision. Vision is an internal mechanism which confirms the creator duly fit humanly to carry out a set purpose. Purpose says, this is what I want to accomplish; vision says, let's go; where there's a vision, there is a commission. Vision is a spiritual feeling of authority, conferred upon the creator.

A creator must realize that his gift is unique. The day he develops his vision is the day of authority. This authority is his power to perform and bring something into existence for the sake and development of humanity. Vision is reality, not an idle dream. A vision can be lost when it is not used. Visions grow as they are worked on. Visions have the tendency to replicate or multiply each other. A vision is still at infancy stage until developed. A vision is a seed, and you the farmer; you decide what you do with it. Vision has its time for execution.

SUMMARY

1. Inspiration and imagination are often occurrences of personal experience.

2. Inspiration is defined as the stimulation of the mind or emotions to a high level of feeling or activity
3. 1st Law of inspiration states that creativity is the product of inspiration and knowledge; pmakumatically expressed as; C=Is x Kn.
4. Imagination is the process of producing and keeping the mental picture in the mind for a reasonable time.
5. Every inspiration is guided by a muse; either positive influence muse or negative influence muse.
6. P.M.Aku's Law on imagination states that imagination is dependent on the strength of the desire of the projected image by the mind of the creator; Pmakumatically expressed as;

   $\frac{SII}{D} \times 1{,}000\%\ IE = IM$ or $\frac{SI^2}{D} X\ 1{,}000\% IE = \underline{IM}$
7. Imagination is divided into (a) keen (b) warm (c) Hot
8. Keen imagination is also called normal. It consists of fragments of mental picture as it occurs.
9. Warm imagination is also called fast or signal imagination. It consists of noticeable passion. Here, confidence becomes rooted.
10. Hot or restless imagination is the point of highest energy, often characterized by restlessness, and the greatest point of realization occurs.
11. Purpose may be defined as the goal, or aim, or motive behind a desire.
12. Purpose may be classified into (a) Initial (b) Defined (c) Established.
13. Initial or half-clear or unsure purpose is not complete.
14. Defined purpose is the point of thoughtful clarity.
15. Established purpose is the point of creative finality, when it comes into existence.
16. P. M. Aku's law of purpose states that an established purpose is the result of the sum of the initial purpose and defined purpose, and without these, has no existence;

Pmakumatically expressed as:

$\frac{I.P + D.P}{Imagination}$ E.P; or

I.P + D.P E.P.
IM

17. Established purpose is classified into (a) Real (b) False or fake
18. Factors of purpose include: respect for humanity, direction finder, confidence booster etc.
19. Problems may be defined in three kinds
    (a) Short term e.g. hunger etc. (b) Medium term e.g. debts etc. (c) Long term e.g financial security.
20. Enemies of Purpose include; (a) Yourself (b) Distractions (c) Negativity and Doubts (d) Cynical remarks.
21. Vision may be defined as an internal feeling of authority in the imaginative strength of a person.
22. A vision is considered a seed which creative potential is subject to the creator who is considered the farmer.

The writer and the composer
Are one;
One writes to warn, the other
Sings to warm
The hearts of those who listen
To their horn.

CHAPTER 14

STYLE

Aims/ Objectives;

1. To teach the student the meaning of style.
2. To expose and expound the types of styles applicable in creativity.
3. To prove that style is a necessary ingredient in the expression of creative works.
4. To appreciate the strength of variety expressed through style.
5. To show that originality is better understood by unique styles.

Style:

By the time you have read up to this chapter, I believe you must be ready to appreciate the way and patterns by which creators would choose, or decide to express their works. I am not a painter but at least I sure do understand that primary colours and secondary colours do exist. In order to express the extent of maturity, I also do know that colours have meanings attached to them, for instance, the colour black would indicate darkness, night hour, white would sing the songs of peace; while green would tell every one of freshness and abundance-say; progress. The musician sure knows his sounds of brass; tone of bass and the lines of the instrumental keys are no mystery to him. These are styles. The architect knows what measurement the room size should bear and the design of the building structure sleeps quietly in his mind. That one is also a style.

The "Creative speaker" in my terms or what is known as an orator sure knows what lines of sentences, its pitch; its effects should be used appropriately with all its body gestures.

Well, good for them and better still for me, if I know where my position on the plat-from of life should be. Thank goodness, I am a creator and a writer. So, you are reading what I write. Call me a stylist and call it style. The creator is not a "mugu" who follows a leading method simply on the grounds of its popularity or blind obedience to its formats, but a guru. The essence of style is to bring positive change and impact through innovations, style makes for variety; style creates room for accommodation of many; style promotes uniqueness and shows the flavours of a unique mind.

My oracle defines the word 'style' as "

1. The way in which something is said, done, expressed, or performed
2. Sort; type.
3. Individuality expressed in One's actions and tastes,
4. Elegance
5. The fashion of the moment b) A. particular fashion.
6. A customary manner of presenting printed material, including usage, punctuation and spelling….to call or name; designate….To arrange or design",

The word "stylist" is defined as "1. A writer or speaker who cultivates an artful literary style". 2. A designer of, or consultant on styles.

Experience has trained me to perceive points of innovativeness in writing and other aspects of life in which I have had opportunities, large or small to work. I have monitored comprehensively and to a good extent, the issue of style in creativity. Herein, I shall discuss my results. This is in regards to writing but I can bet you that if you keep an eagle-eyed on certain issues and read carefully, you might find some where in between the lines a statement; a principle which relates to the other aspects of creativity.

Generally, style whether of fashion, paintings or writing falls into two categories;

1. Stagnant style 2. Progressive style. 1. Stagnant style: this is the style which has a redundant movement pattern. The stylist hates and resists any change to his former way, He is an addicted-a he rent to his ways of doing things. He is not innovative in culture.
2. Progressive style: This is also known as Innovative style. The stylist moves with time. His mind is active. When you delight in his work

of yesterday, he laughs at you because he is a thousand nautical miles ahead of you already. These are the designers of the modern world and humanity would not be better without them.

The writer

Though a creator of developmental programs, I began from the art of writing. I am a writer, I hereby, do define style, as specific to writing in my private discovery over the ages as.

Modern writing style: This is a also referred to as Longitudinal or Lines-stretch (L.S.) style.

It comprises the conventional method of school writing, story telling, along the full stretch exercise book. Sentences are stated in full to show the parts (Subject, Object, Predicate). Rhymes are not necessary but meaning.

Example:

"Since Johnson visited our home, we have been thinking about his last speech," said Tunde. Personally, this style does not please me much.

1. **Shakespeare writing style:**

   I shall humbly refer to this style as it pertains to my works as Akustick style. Here, Statements are broken into stick-bits or stanzas. With an overhelming influence in the plays written during the days of William Shakespeare or Shakespearean days, the object of the message lies powerfully, in the use of rhymes; a play on words.

   Rhymes or sticks (with reference to broken sticks) are respected and the prowess of words storage and usage at the writer's disposal is well defined. This style favours me, apart from the reasons stated above, and as well creates opportunities for me to bump or create a word spontaneously without thinking; yet having it fall into place in the right space. For instance, "slappeido"; and "fallacious-decaffeinated ant "many" naturally occurred to me as "pimps" or words in my play; "Valley of the kings".

   Words also occur to one who is vast in reading and doing a lot of practice in writing. Its greatest conveyer is Emotion or passion.

Example:
Screw the Jack,
Taught-teacher Jack
It's my new tactis
Found in mathematics.
Style defines your personality and speaks volumes of your uniqueness.

Advantages/ Merits of Creative style

1. To promote innovative creations
2. To show excellence in ability expressed
3. To allows for accommodation, i.e. doing the same in many more interesting formats.
4. It helps for originality.
5. It shows the strength of variety in creativity.
6. It adds value to creative works.
7. Better understanding of works and its origins.
8. It honours the as creators
9. Diversity is appreciated
10. Beauty is the satisfaction-its end product.

Style also defines a writer uniqueness through his use of language and his selection of words. When the ancient language and modern style is combined in a literary work, I refer to this as complexity; for instance, the use of an African proverb in an urban setting instead of rural, or, using the Shakespearean style in a traditional African plot and setting. When meanings are conveyed in certain rigid manner, especially in a play, I shall refer to them as simplistic.

Well, your style is your definition and this is my position.

Style also appreciates culture and its changes. The creator must, as a stylist, have a high "Predictability Ratio Content "(PRC) i.e. the ability to predict or foretell certain ends of behavior or character. He must be physically and spiritually balanced; not substituting fasting for hunger's nor prayer for Labour; must be provoking and persuasive, attacking and defending at the same time. After all, who is a creator? One whose life is of inner search and continuous understanding.

THE AKUSTICK STYLE OF WRITING

The term "Akustick" style of writing, as explained earlier is a term derived from its author and developer; Peter Matthews Akukalia; also known as P. M. Aku. The combination of this abbreviative would be Pmaku (i.e. pronounced as Maku). It was developed and first used by its author in his work titled "stranger than strange" in which its genre of literature was called the propoplay. Though first written in the year 2010, April; it was published a year later.

Nature of the Akustick style

The nature of the Akustic style or method of writing is such that welcomes its inclusion into the play and poetry exclusively. It cannot be used in the prose or hovel since statements are often broken into parts known as sticks, and a prose would require whole statements in order to drive home a complete meaning to its reader and the student. This is known as exclusive accommodation from this, we may rightly state the kinds of exclusive accommodation as:

i. Play or Dramatic Exclusive Accommodation
ii. Poetry or poemic exclusive accommodation
iii. Play or dramatic exclusive accommodation.

1. <u>Play or Dramatic Exclusive Accommodation</u>

Here, the Akustic style is used throughout or sometimes partly to write a play, propoplay, poetry or nidrapoe.
Its characteristic features include:

i. Separated character placement (SCP):

Here, the character title is separated a line ahead try the spoken speech.
An illustration could be:
John;

It would matter nothing to
us at all to do things accordingly;
nevertheless considerations
should be made much sensibly.

Rather than let it follow up in the same line of arrangement, it would be made into distinct yet adjoined parts.

ii. Spacing:

This is noticeably, a unique spacing format,especially those between two characters after speeches. See below:

John:

It's not fair to run errands
For those who little appreciate.......

Susan:

Prove thou thy statement further
condemn no one on stands so hard.

iii. Centre spread format

The characters, though placed at the page starting line, its attendant speeches do not start from there, but are spread out on the centre page. This is called the centre spread format.

2. <u>Poetry or poemic exclusive accommodation.</u>

Here the Akustic style is used to write poetry or poemic genres such as the nidrapoe. It is functional if it uses the centre spread format; and disfuntional if it does not. However, this is not exceptional to poetry alone but to play and propoplay alike.

Classes / kinds of Akustic style.

The classes /kinds of Akustic styled methods of writing or word artistry is explained thus:

1. longitudinal class:

Here, words are carved or written according to a string-length design. Below is an example:

Were it not a better yardstick to
Be measured from a farmer's sickle;
When a greater harm it would be
To be known and studied of medicine.

2. Triangle style:

This is the use of the triangle design to express words:

Poor man; or poorly married

Should he not have fully
Given thought to whom he loved
For money and marriage duly
Live this day in their own misery.

3. **Pyramid style:**

Often a long expressed group of statements,
It develops fully from the triangle style. Its minimum is often seven statements.

4. **Tit-bits style:**

These are statements or speeches expressed from the long and short format. This could be seen below:

I hate the very earth; coward
Without an apology
Which you ever thread upon
To humanity and history
Bids you farewell in every game
You ever shall play
If thou art all, shall play.

5. **Combined style:**

This is the use of the known style e.g. longitudinal; and the unknown or unpopular e.g. straight line; slantyness style; as measured on the line design from visible view.

PRESERVING CREATIVE RECORDS AND CULTURE.

Preservation of creative records cannot be underestimated going by its importance, role and psychological impact it has on individuals and the larger society. It is necessary to state here that the keeping of creative records, is as important as keeping an organisation's records (book keeping) or deeds of certain person's in politics and other fields (history),keeping of a person's life records (biography),and of course the very popular of them all, the library. The preservation of creative records goes a ling way down in history, as far back as the creation of man and its pre-historic existence, the creation of the holy scriptures and books; special events and their era of existence.
From these, we may deduce the kinds of creative records intuitively. These are: intellectual records, monumental records and celebrity profile.

1. INTELLECTUAL RECORDS:

These are records of high academic standards and creations. They involve rationalism beyond the lessons of the classroom. They often define loudable brilliance and sustainable intelligence. It is often proved that certain people who have made exceptional intellectual creations such as in literature, science and perhaps, propose magnificent and workable management systems have at one time or the other had direct or indirect academic problems. Yet, one wonders how it is that they have excelled exceptionally at creativity. The answer is simply at the application of local intelligence, what we today know as common sense.

"Further proof to this would be in the following excepts;

Following the death of his wife in 1885, Thomas Edison began to spend his winters in Fort Myers. Deciding that he liked the quiet river front town, he had a house built to his specifications in Maine and shipped by schooner to Fort Myers, where the pieces were assembled. Today the unpretensious main home, guest house, and laboratory are open to the public. Among several of Edison's innovations in the main hose is a lighting system, installed in the 1880's, using handmade light bulbs. Edison's laboratory is filled with test tube and other apparatus. It was here that the tireless inventor conducted experiments in his quest to produce synthetic rubber. The tropical gardens are lit at night with power supplied by underground wires that Edison installed. The adjoining property was owned by Henry Ford, and the families wintered together until Edison's death in 1931.The Ford home is also open to the public. Located at 2350 McGregor Blvd. in Fort Myers."

-Great American Journeys no 6; page 41.

Records such as these however would not have been possible to reach or find if they had not been preserved over the years. The importance of preserving creative records is thereby proved.

An understandable fact of these people is that they have often applied the "self development strategy", often out of a high interest level; what we may refer to as "creative records strategy ". These people are an example of those whom my friend, Dr. Oshionebo says "do not need to go to school because we study what they create. The self development strategy (SDS) or creative records study (CRS) has its greatest rooting or sourcing from private reading. It is illustrated below:

PRIVATE READING CREATIVE RECORDS STUDY
SELF DEVELOPMENT STRATEGY.

Either born out of the desperate need to survive, excel or achieve;
One thing is clearly proved here; that the psychological content of an individual's mind is determined by the content of the stuff he reads more often, which has its overall impact on the society. This is what I may term the Law of Self Development.

Let us comprehensively study the following notes briefly;
"at the base of sustainable development is continuous learning. The very foundation of continuous learning is a little known art called Reading. If the potency of voracious reading has not caught up on you, then you haven't really known the following persons: Thomas Edison, Michael Faraday, Peter J. Daniels, Phillip Emeagwali, Ben Carson, Sunny Ojeagbase and Kevin Callan."(Guardian Newspapers, Tuesday March 29, 2011. pages 36,37).It is important that such creative records are inculcated into our education curricular. This would help the child and student imbibe the right moral al values found in the creative culture.

2. *MONUMENTAL RECORDS.*

These are records of great personalities which are kept in museums for reflections, statue representations, specific naming; perhaps of streets, states, bridges, airports and localities. Does it not interest you that the city of Pennsylvania was chartered in 1681 by the work of William Penn and there after named after him. Interestingly, monumental records have long been part of our culture; but if it is given due recognition, then predictably a new lifestyle of honour due every work and person of worth would have been developed. By the creation of museums, employment would have been created and life would be more meaningful.

3. *CELEBRITY PROFILE.*

This form of creative record keeping ensures a stay of achievements made efforts which do not bear a link to either intellectual or monumental. It is the records of bagged titles, awards from sectors such as sports, music, comedy etc. These would include records such as the accounts found in the popular Guinness Book of Records, records of the late king of pop, Michael Jackson, etc. However, a person, intellectual and witty enough may be able to excel through the accounts successfully.

BENEFITS OF CREATIVE RECORDS KEEPING

1. For preservation purposes.
2. To act as a source of inspiration to others.
3. It proves the existence of opportunities for work.
4. It eliminates the mediocre mentality in people.
5. To push for the achievement of excellence and nothing less.
6. A possible source of employment and wealth creation.
7. Such records can help us monitor the rate at which the generations, time, era, and a people have progressed.
8. May act as aid to studying human behaviour and culture.

SUMMARY:

1. Style is defined as the method of expressing a creative work.
2. There is style in every aspect of creativity.
3. Categories of style are:
   Stag rant progressive
4. Stagnant style is characterized by a redundant movement pattern; with a clear resistance to change.
5. Progressive save style is also referred to as Innovative style. Characterized by constancy with time and open mindedness.
6. The writer is a creator
7. Writing way be classified in to;
   a. Modern writing style or longitudinal or lines-stretch.
   b. Shake's pearean writing style or Akustic style.
8. The greatest conveyer of creative writing is emotion or passion,
9. Advantages of creative style includes; promotion of innovation, excellence in ability, allowance for accusation, originality, variety, value, under-standing of origins of works, honor to it's originator, diversity appreciation and beauty.
10. The creator must have high predictability ratio content i.e. the ability to foretell the end of certain behaviour or character.

CHAPTER 15

THE CREATOR'S THEATRE

Aims /Object

1. To teach that every creative process takes place, and ought to have a place for the activity of creating.
2. To teach that a conducive environment is a necessary factor for creating.
3. To show that a theatre is, and should be recognized as a place for creating and so accorded every respect for that purpose.

THE CREATOR'S THEATRE

As discussed earlier in the preceding pages, the places of creativity, which encompasses all forms whether of the sciences, Arts or the commercials background, of any sort or kind, not discriminating of gender or race; at any place or time, should be referred to as a theatre. If a doctor's sacred place of treatment such as surgery room or delivery room is called a theatre; the place or stage of acting plays a theatre also; it is most acceptable to conceive and speak of the creator's theatre.

The theatre generally has certain conditions of existence. These conditions are made so as to achieve total reverence credibility, and focus by whoever is using it. The conditions of a theatre are stated below:

1. It must be devoid of noise
2. It must be orderly arranged; book for shelf (often to be called a private study or library); equipment for the arrangement (such as pens, pencils, papers, materials of design).

3. The theatre must be free from intruders or interrupting discussions, gossips, hosting of people exception very serious issues as allowed by the creator.
4. The theatre must be considered a place of sacredness where one communes with his creator (Prayer) communes with his treston (imagination, inspiration and pictural understanding); communes with his intelligence (academic works e.g. reading and study, research and development).
5. It must be considered his first office since this is the laboratory for which he brings things to reality.
6. It must be clean, well ventilated and tidy since "Cleanliness is next to godliness". A dirty environment promotes an unruly mindset.
7. It must be kept away from undue children interference.

The creator's theatre must be one of an inestimable value to the creator, and so must be well ordered in all of its essence. The theatre must be considered in all of it's essence. The theatre must be considered the centre of the creator's activities and so accorded every due respect.

SUMMARY

1. Any place at all reserved for the purpose of creativity, is a theatre.
2. The conditions of a theatre should be devoid of noise, orderly arranged, equipped with the necessary materials, free from intruders, considered sacred or private, well ventilated, treated as an office, clean and tidy and kept away from undue children interference.
3. The theatre is the centre of the creator's creative activities.

CHAPTER 16

CREATIVITY & IDEAS DEVELOPMENT

AIMS/OBJECTIVES:

1. To teach that the creative person is a raydealist by philosophy.
2. The philosophy of raydealism appreciates the attitude of evaluating a situation as present, working for better times, with an optimistic view of tomorrow.
3. The combined philosophies of realism and idealism support the pure philosophy of the creator.
4. To teach a proper method of viewing ideas.

Creativity & Ideas development

It has been said that the world is enjoyed by heroes and heroes are those who develop ideas, design it and bring it to reality. The creator is a combinative factor of the philosophies of realism and idealism. An idea is a concept and every concept is an idea. First, let us define the terms mentioned above and try to see how we may find their connecting points in order to maximize our potentials as found in every idea:

1. Realism.

The American Heritage dictionary defines it as; "An inclination toward objective truth and pragmatism. The representation in art or literature of objects, actions, or social conditions as they actually are". Such person of this school of thought is called a realist.

The creator works at things, reality and conditions as they are and searches out truths, value, beauty and profitability from them. He does not deny the facts of painful realities such as hunger, poverty, unemployment, man-made disasters such as war; nor natural disasters such as earthquakes, or volcanoes. Rather, he would make a story and its lessons out of it (novelist or

essayist) poem (poet); funny pictorial remarks or representation (cartoonist or Drawer) carvings (sculpturist); paintings (artist) e.t.c. His essence is to take positive advances and search the hidden opportunity in such potential He obeys the first law of a true creative realist which I shall state.
1st Law of creative realism states that truth to one self is the greatest pillar to one's conscience; this pillar falls the day truth decays.

It is therefore imperative that we understand most objectively, by the guide of our inner mind, the reality of every situation. It is therefore a wrong attitude to judge issues based on mere appearance or inadequate facts or opinions and views.

It is important, however, to state that the word realism comes from the word "reality" meaning, "the quality or state of being actual or true. One that exists objectively; or perhaps the word realize which connotes, "to comprehend completely or correctly; to make real; fulfill; to obtain or achieve as gain or profit". This must be a guiding philosophy of every creator or creative scientist.

2. Idealism:
This is the philosophy or theory that things, in the selves or as perceived, consist of ideas. In itself, idealism is the practicem of envisioning things in an ideal form. Also, it is described as the pursuit of one's ideal. Further more, derived it is from the word "Ideal" which is a concept of something as perfect; a standard of perfection or excellence; an ultimate objective; goal; an honorable or worthy principle. The word has its principal derivation fro the word, "IDEA". The word" Idea" would mean vividly: something, such as a thought, that exists in the mind as a product of mental activity; an opinion, conviction, or principle; a plan, scheme, or method, a general meaning or purport."

But compare this with the word "Strategy" which means," a plan of action; the art or skill of using stratagems-which in itself means "a scheme for achieving an objective".

The creator is an idealist in his own very nature. He lives by ideas, sleeps in ideas, feeds ideas and feeds on ideas, marries ideas and keeps society by sensible ideas. He is a creative idealist. By this, the philosophy of the creative scientist shall be referred to as the philosophy of Raydealism, which is a combination of the words "rea-from realism, idea-idealism; and 'lism from both found in both words, It therefore denotes that in principle, the creator or creative scientist is operated mentally by the philosophy of raydealism (pronounced as ray-dea-lism). A raydealist, not definitively, is a person who sees things the. May they are, appraises them, envisions a better, intelligent attitude to it and then works towards its achievement. He does not take chances and does not give either, whether exclusively (by excuse of any kind) or complainingly (by complains). He does his best at all times and takes responsibility for his own decided actions. In order to understand how we must, it is however necessary to refer to Robert schuller, in his book, 'Tough times Never Last, But tough people Do "in relation to valuing our ideas, for by ideas are great nations built and famous wealths made. According to him, "There is something wrong with every good idea. Any time God gives you an idea, you can find some negative aspect to it. It's amazing how people sit in a deliberating meeting and respond to an opportunity only by finding fault with it Don't throw a way a suggestion when you see a problem-Instead, isolate from the possibility. Neutralize the negative. Exploit the possibility, and sublimate the negative. Don't ever let negatives kill the positive potential that is within an opportunity. Nothing is impossible if I will hold or to the idea that it might become possible somehow, someway, with someone's help". Then he goes on to add; "Never reject an idea because:

You won't get the credit;
It's impossible;
Your mind is already made up;
It's illegal-you should never violate the law....you might able to get the law changed! You don't have the money, manpower, muscle, or months to achieve it!
It will create conflict;
It's not your way of doing things!
It, might fail;
It's sure to succeed."

SUMMARY

1. Every work of creativity is powered by ideas.
2. Realism is a conventional philosophy which evaluates situations as they are.
3. Idealism is a philosophy founded on the fact that things in them selves consist of ideas and thinking about their existence in an ideal form.
4. Radicalism is a combined factor of realism and idealism.
5. The study of creativity is guided by the philosophy of raydealism.
6. The creator or creative scientist is by conviction a raydealist.

CHAPTER 17

CREATIVITY & EVALUATION

Aims/Objective

1. To teach the students the various methods of developing reasonable, sensible, logical and profitable ideas and creative works.
2. To teach the students the applicable steps to evaluating their ideas and works for profitable use.
3. To impart into the student the importance and precise methods of expressing and communicating their various motives, purposes and lessons behind their creative works.
4. Through this medium, the lessons of copyrighting by way of protecting one's works is learnt.

CREATIVITY & EVALUATION

Creativity & Evaluation

Creativity is born of works. While every one is endowed with a creative average, the possibility of performing outstandingly or creating outstanding works depends on ability; and more on decision and determination.

The creator's steps to creative evaluations of his works are as follows;

1. Clarify the Design specifications & constraints:

Describe the problem clearly, definitively and realistically. Make notes of these for easy recollections noting constraints and specifications. Set the time limit, guidelines and principles for the design you are creating or the problem you are solving with that design.

2. Private Research:
Book, materials, and previous works similar in nature and scope must be investigated, previous creator's background must be known in order to gather skill and knowledge for the work.

3. Create Alternative Designs:
The presence of options in creative thinking or mental hazzles is important. It aids for advance reasoning, innovative approach; responsible logic and critical appraisal of details. Create many other ways of doing one thing or performing the same task so as to chose by evaluation, the best; not too flashy, not too dull either (if it were in graphics).

4. Make a choice:
Choose among the variables, the design which best appeals to your specifications, fits within the constraints, and eliminates more of its characteristics.

5. Design a prototype:
Bring your mental picture to its first work. This, I shall refer to as the principle of creative initial; whereby a work of design is first made in order to bring to survey every detail which has led to the creation of that work. This principle gives room for scrutiny, change of positions, replacement and addition of necessaries and its over all impact on you, in terms of time spent and resources used.

6. Test and Evaluate:
How will you know whether your design will meet specifications and solve the problem? Thought testing, you will document what improvement to make and why.

7. Redesign the solution:
By a careful and objective comparison of previous works, make upon your opportunity for improvement perhaps in speed (ma chine); volume (Books & literary); clarity (paintings and sculpture) enhancement (business and general techniques.)

8. Communicate your achievements:

This will include total expressiveness in terms of work presentation, methodology, your experience and the reason of your choices, and then your next point of improvement, since the creator always has work to do.

Creativity analysis & evaluation:

This page is the point of records whereby the activities of the creative scientist is recorded, investigated, analyzed and observed. It is the result of the practice made by the creator.

During examinations, it must be seen as the practical point. Its result structure is accompanied by the P3 of the SCAP and general academic results structure.

Creativity analysis Sheet (Cash)
Personality Report.

1. Name: ______________________________
2. Date of Birth: ______________________________
3. Year of Admission: ______________________________
   Works Report: ______________________________
4. Name of Work: ______________________________
5. Nature of work tick: Literary Technical Li-Tech
6. Purpose (s) of work: ______________________________

7. Please state your motivation: ______________________________

8. If innovation, state previous creator, life years and works e.g. John Tinkin1880-2012; and date of invention.

9. List materials you used during this work: ____________________

10. Did your final design meet the initial specifications! ____________________

11. If you had to redesign your solution, what problems did you encounter? ____________________

12. By simple tape rule, what is the measurement of your work i.e. length and breadth; how-long did it take you to complete the work?
    A. ____________________
    B. ____________________

13. Describe how the invention affects human safety or comfort.

14. Describe the desirable or undesirable results produced by the work

15. State the design challenge you encountered ____________________

16. State what you have learnt from your experience ____________________

17. State your message's key ideas e.g. literary to increase reading culture, political work to aid government in its overview of policies through cartoon; textiles etc. ____________________

18. State your audience / market

19. Is your work copyrighted or registered? State details.
    Tick Yes No Awaiting [with proofs of filled form]
20. State your expectations for this work:

NB: This "CASH" instrument is strictly for professional use.

SUMMARY

1. It is necessary for the creator and creative scientist to be guided by the guidelines behind every creativity.
2. A creative work without proper expression may likely not reach the appropriate audience.
3. Every complete creative work is a couple of ideas.
4. It is necessary that the key ideas behind an original work is defined by the mind of the creator
5. Documentation for the records and further transmission of generational knowledge is the greatest priority for every completed work.
6. .
7. Every complete creative work is a couple of ideas.
8. It is necessary that the key ideas behind an original work is defined by the mind of the creator
9. Documentation for the records and further transmission of generational knowledge is the greatest priority for every completed work.

When the pen comes, is the fire
making our minds soar higher
wielding the sword is the writer
scribbling his mind upon the paper.

To the writer, words are food
showing how much he would
dissect every truth in the hood
proving enough to form a book.

Words are food properly cooked
sharp traveling arrows;
words are chokes for the crook
bending low as the bow.

CHAPTER 18

NEW!
THE PROPOPLAY & NIDRAPOE

Aims/ Objectives:

1. To introduce the student to the newly created genres of literature.
2. To prove that creativity may be applied to existnig subjects, works etc; in order to create innovation.
3. To prove that originality can be created from a previous original.
4. To prove the dynamism of the mind when applied to improving knowledge and excitement.
5. To teach that every original work, when fully developed, is practicable and profitable.
6. To prove that bounteous opportunities still exist for further human excitement, dignity and development.
7. To prove that there is no limit to creativity since knowledge forms the basic foundation.
8. To reveal the greator Joys which emanate from the recognition that one is credited for making positive impacts through his works.

NEW!
THE PROPOPLAY & & NIDRAPOE

Where progress is desired, change is inevitable. A nation without a prophet is doomed so quoted the wise king Solomon. Here, I add that a people without a creator is lost. We do often fail to realize that our heritage is a product of our own creativity. This is a study and no study exists without a variety; if so, it is not worth, in practical sense to be called a study; fragments but fitting into a whole.

The propoplay? The Nidrapoe? What is that? Strange words you would say. Wouldn't it be insensible to blame a creator for creating? The truth of the matter is that everything happens to be creations.

Through this medium, I shall introduce myself officially much beyond the titles of a play wright, a poet or a novelist, a private researcher, a creator and a facilitator, those are common terms to the preset world, and would remain beyond the next millennium, that is one thing I am sure of. The truth of the matter is that I hate anything considered objects of common place or too popular to be perceived with the eyes of triviality. Humbly, I will like to be called a creator in addition to other issues; drapoet (pronounced Drapoet) and prop playwright or peter writ; your Excellency.

The essence of this work is to promote the creative potentials behind those matters considered insignificant, which when harnessed would add value to our lives.

I have during the course of my personal research discovered these hidden factors which are stated here in, and I hope to discuss the pending matters here in. Too many stories!

WHAT IS A NIDRAPOE

The word "Nidrapoe"; was coined from three words: Ni-Nigerian; DRA-Drama: POE-Poetry. It Means "drama written within a poem or poetical format "or "a poem written to tell a story or narrative "with a special source from myself as a Nigerian. For better explanation, I would recommend the book: "Glimpses in verse-a festival of poetry: (containing 250 poems)" see the poems on "The old man's lane (Part I II III)" Evening at my Balcony" etc.; in order to appreciate the nidroapoe. The person who writes a nidrapoe, I shall call the drapoet i.e. the drama-poet. As simple as this definition may sound, it take a considerable effort to write this aspect of poetry. Anyway, poetry has naver been easy either.

FEATURES OF A NIDRAPOE:

1. Projected to become the most difficult form of poetry.

2. Combination of characters and poetic devices.
3. Based on a segmented and voluminous system of poetry.
4. A form of poetry often told as a narrative.
5. Characters may be bracketed or underlined
6. Contains no scenes nor acts but chapters and marked verses.
7. The strength of a Nidrapoe is measured by the number of verses and rhythmical tenacity.
8. The message of a Nidrapoe is often connoted by its title.

The first complete work on a Nidrapoe is titled "Love's a fever" with over 800 verses, and "The moon Walker" with 1,500 verses and was done by the creator, and author of the field himself; peter Matthews-Akukalia. The strength of a Nidrapoe proves the extent to which the drapoet knows his subject. Firstly, it discusses only a topic selected as the title and a book is formed from there. This implies that poor understanding of the issue discussed would expose the drapoet to ridicule. Here, verses and chapters are numbered, continuously setting new limits for records. Convincingly, the nidrapoe is a sign of new things to come.

Kinds of Nidrapoe

1. Fictional Nidrapoe: This is a kind of nidrapoe which is based on a non-living person, and is often based on abstract issues e.g. Love's A fever; which contains about verses and 847 over 10 chapters.
2. Biographical Nidrapoe, this is a kind of nidrapoe which is based on the lives of persons either dead or alive e.g. The Moonwalker which contains 10 chapters and, 1,500 verses on the legendary Michael Jackson.

WHAT IS A PROPOPLAY?

The word 'propoplay was coined from the existing genres of Literature which are Pro-from prose; Po-from poetry; and play-from play; This genre of literature is what in my most honest intention, the combination of these three in a book. They exist together, telling the same story in different roles and parts, yet being knitted together towards the end goal of producing a complete story. It goes with a title and then the inscription-a propoplay. I believe this would bring a new dimension to literary works. Indications to mark the next section are made e.g. "Now story begins", or "The Drama", the poetical aspect may either be combined or mixed in between Acts or be left separately.

Kinds of Propoplay

I shall divide the propoplay into four types or categories:

1. Prose-poetry-play combination e.g. "Stranger than stranger" by Peter

Matthews-Akukalia.

1. Prose-poetry-play combination e.g. stranger than strange (the angel story).
2. Play-prose-poetry combination.
3. Poetry-Play-Poetry combination.
4. Prose-poetry-play-Mixed combination.
5. For the above, as the novel story is told, there is a sudden intrusion of poetry or play within; just fitting itself into the right frame. See an example below;

"As time passed by him, he felt astonished at the magnificent tunes played at the inner room. He stood up and walked to the window to view the other side of the room when Daniel said to him;

standing! Standing!
do now no more unpaid
job of a policeman; as
you view the roads, they
view you also to no gain.'

Stage attitude

The stage attitude (Stat) defines the methods or stage craft with which a play may be acted. Apart from the discerning ability of a director, or the presentation made in the book, the whole prece of the propoplay is actable except that for the prose aspect, the narrator or teller or reader recites few necessary lines behind curtains.

Benefits of a propoplay

The benefits of writing and reading a propoplay can be summarized below as;

1. Combined satisfaction: which comes from enjoying all genres of literature in one book.
2. Imaginative Prowess: Here, the imaginative prowess its extent and ability as endowed in the writer or reader is tested and developed for strength.
3. The different aspects of the propoplay may be enjoyed distinctly and collectively.
4. The same satisfaction may be derived from stage work and literary work simultaneonely.
5. Accommodation: A propoplay exists in variety and kinds which gives room for the writer to choose a suitable style of writing.
6. There is a brief introduction page for each stage of writing e.g. "Now the story begins "or "It's story Time", "poetry of the heat", then the Drama Begins".
7. It could be a long poetry with a short or long prose or play or vice versa.

   A person who writes a propoplay, I shall refer to as a **propoplaywright or Peterwrit.**

The guiding principle which has led me to the discovery and development of both forms of literature is what I shall refer to as **"The combination theory or philosophy".**

The P.M. Akus combination theory or philosophy states that:

Unity can be created from the separate or district existence of an Original work, itself a creation of originality." i.e.

+ = or

Types of Propoplay:

1. Fictional propoplay: This is a propoplay which deals with issues of imagination and are often abstract in nature. They are inanimate e.g. stranger than strange (hours of judgment) by Peter Matthews-Akukalia.
2. Biographical Propoplay: This is a kind of propoplay or peter writ based on a living or dead person.

CHAPTER 19

INSTRUMENTS USED IN CREATIVE SCIENCES

Title: The Pmakumeter Guage.
Contents: Contains the calibrated curiosity levels table. It's title Pmakuneter Guage is derived from the inventor's name Peter Mathews Akukalia (also known as P.M.Aku. He is the creator and father of the field known as the pimples of creative science (pronounced as Pmakumeter;) slent psound

Purpose; It measures the energy level (upsurge and down surge of curiostory measured in curiository Joules (CJ) indicating the point of Increase (PT) and Paint of Decrease (PD); the levels according to the peak of Aspiration (POA); the categories of the thermometric thinking level (TTL) which is obtained after due calculations to determine the individual's level of interest.

Material/ Nature: The Pmakuweter Guage is made in three formats depending on the use, and class of purpose.

1. Glass 2.Transparent plastic 3. Wood it is a compulsory instrument for this course as it light vital for it's purpose.
2. Title: Creativity Analysis sheet. (CASH)

Purpose: It was designed to help the creator understand better the steps and various analytical steps required to develop and evaluate his ideas techniques and works.

Contents: it is as important to the student of this course as the ledger is to the accountant. It is personal to the student's use. It is vital for tests. and examinations. A creative scientist or student's examinations on this course must be practically appraised by his physical work (s) and statements professional examinations to the genius stage, sage stage and the chartered sage stage are judged by the creation more outstanding works. It is outstanding when considered unique and devoid of any question ability as regards to cheating, initiation and other vices. It is also important that the student explains his/her works. A works. A work is audited to cash. The authorities, approved by the guiding regulations must keep there records and an well copies made and kept by the student or scientist.

However, it's important to note that while the general study of every student is necessary, this does not quality their to be called creative scientist after attending to the professional qualification of either the master/mistress; Genius/Mistress De Genius; Safe/Misfiressde & chartered sage Ch.s

Drary of Ideas (DI)
This is explained earlier, is the bark material meant to jot down initial ideas, observable trends which might seem to make some sense later in the future. Its rules are stated there in-copyrighted

3. Pictorial representation of the mind Dichotomy calendar formal-used for illustrative purposes copyrighted.
4. Pictorial representation of the structure of the creative mind (calendar format) wed for illustrative purposes.
5. Plastic copyrighted format of the structure of the creative mind patented-for illustrative purposes.
6. Fuseable plastic format of the mind dich-to my; often separable by joints- patented.

Rules for the use of the creative laboratory instruments.

1. They must constitute the materials present at the creative sciences laboratory/ theatre.
2. They must be used applicable when their specific topics are to be taught.
3. The Diary of ideas, Pmakumeter guage and the creativity analysis sheet are to be compulsory purchased for personal use of the creative scientist during private and class studies.
4. The personal ownership of the other illustrative instruments and materials is optional.

5. All materials are essential for practical and theoretical examinations on the principles of creative sciences.
6. The Pmakumeter guage and Cash must be individually used, accompanied by copy of work during examinations.

Benefits of the instruments of creativity.

1. Aids the student to be adaptive by practice
2. aids the process of imaginative thinking.
3. Develops in the student the ability to constructive and logical reasoning.
4. Helps the student make sensible comparisout and deductions.
5. Develops the stolen towards the appreciation of periodic reflections.
6. Necessitates the importance of work review.
7. Assesses the student's innovative ability to develop new models of previous works.
8. Builds confidence and positivism.
9. Helps to defines true belief thought faith and other virtues.
10. Proves the relationships between decisions and responsibility; its personal and collective effects.
11. Helps develop creative focus
12. By these, the student would learn to be open, sincere realistic in setting objectives and docile in his apt to study and learn the applicable dynamics of change.

The above qualities are those which build the ideal mind. These collectively lead to success and happiness, firstly in the mental and than conclusively in the societal, measurable and honourable.

5. Mathematics [illegible] encourages practical and theoretical examination on the principles of creative sciences.
6. The fundamental science [illegible] must be [illegible] dealt [illegible] with its [illegible] dimensions.

Benefits of the mathematics to creativity:

1. Aids the student to be adaptive by practice.
2. [illegible] reasoning [illegible] thinking.
3. Develops in the student [illegible] ability to [illegible] reasoning.
4. [illegible] the student makes [illegible] conclusions and deductions.
5. Develops [illegible] towards [illegible] importance of [illegible].
6. [illegible] the [illegible] works.
8. Builds [illegible] and [illegible].
9. Helps [illegible] defines [illegible] virtues.
10. Proves the relations between [illegible] and [illegible].
11. [illegible] creative [illegible].
12. [illegible] the student would [illegible] study and learn [illegible].

[illegible] are those which build the [illegible] collectively [illegible] and happiness [illegible] social [illegible].

SECTION 2:

Creative Education

Practical & Functional
Counseling Tips.
Comprehensive Scholarships.
Success Oriented info
For Universities & Tertiary Institutions.

1. Diary of Ideas (DI)
   This, as explained earlier, is the Bark Material meant to jot down initial ideas, observable trends which night seem to make some sense later in the future. It's rules are stated there in-copyrighted.
2. Pictorial representation of the mind Dichotomy calendar formal-used for illustrative purposes copyrighted.
3. Pictoral representation of the structure of the creative mind (calendar format) used for illustrative purposes.
4. Plastic copyright format of the structure of the creative mind patented-for illustrative purpose.
5. Fuseable plastic format of the mind dichotic my; often separable by joints patented.

Rules for the use of the creative laboratory instruments

1. They must constitute the materials present at the creative sciences laboratory/ theatre.
2. They must be used applicable when their specific topics are to be taught.
3. Thee Diary of Ideas, Pmakumeter guage and the creativity analysis sheet are to be compulsory purchased for personal use of it creative scientist during private and class studies.
4. The personal ownership of the other illustrative instruments and materials is optional.
5. All materials are essential for practical and theoretical examinations on the principles of creative sciences.
6. The Pmakumeter guage and cash must be individually used, accompanied by copy of work during examinations.

Benefits of the instruments of creativity.

1. Helps the student to be adaptive by practice

2. Aids the process of imaginative thinking.
3. Develop in the student the ability to constructive and logical reasoning.
4. Helps the student make sensible comparisour and deductions.
5. Develops the student towards the appreciation of periodic reflections.
6. Necessitates the importance of work review.
7. Assesses the student's innovative ability to develop new models of previous works.
8. Builds confidence and positivism.
9. Helps to define true belief through faith and other virtues.
10. Proves the relationship between decisions and responsibility; it's personal and collective effects.
11. Helps develop creative focus.
12. By these, the student would learn to be open, sincere, realistic in setting objectives and docile in his apt to study and learn the applicable dynamics of change.

The above qualities are those which build the ideal mind. These collectively lead to success and happiness, firstly in the mental and then conclusively in the societal, measurable and honorable.

Education still has a voice
Though silent;
She still has a strength though weak;
She is the posterity of the young;
She is the strength of the professions;
The very foundation of a nation's progress;
She deserves to be heard;
the most deserving of all mother the maker of all valiants.

Supreme knowledge rests in the
hands of him who has discovered
Self, who knows where he is going
Who has unfailing determinate for reaching
The object of his heart's desire,
Who has for mulated a workable plan
For his attainment;

This saying beings only to the
True citizen, the best of patriots.

It's proved globally that only
the citizens of a country
hold the will to drive the wheel
of their country forward;
I believe the only way to achieve
This is to operate the Educational
system as a collective responsibility

All other systems seen to introduce
Innovations continually;
Yet, our educational system ought
To enjoy this facility size it is
The tree which house the fruits.

Where there is lack of scholarship
opportunities;
I doubt if the best brains would
ever be discovered;
Since only the rich would able
To fund education.

The fathers of our freedom were
Products of Educational opportunities;
Why should we not seek for more
of such good opportunities?

Sustain Functional Education consisting
For theories and more practice;
Then breed a nation of practical
And sincere leaders.

There is nothing you offer for
the promotion of intellectual culture

that is enough; it remains everyone's highest
duty for true attainment.

He who fights for Educational
Emancipation;
Must hold the Law firmly.

No Education, no Nation;
No functional Education,
No functional Nation.

Education must lead the pace,
For others to follow;
without this, we become narrow,
and fallen, even by our own burrow,

My Letter
To The Citizens
To The Government

Dear Sir/ Madam, Student

INTRODUCTION TO THE GUIDING PRINCIPLES OF THE EDMP

According to the united Nation Declarations of the Rights of the Child, its Utmost purpose is to promote the freedoms, therein set forth to enable the child to enjoy a happy childhood for his own good. It obligates not only the government but also "parents, upon men and women as individuals, and upon voluntary organizations, local authorities to recognize these rights and strive for their observance by legislative and other measures progressively".

PRINCIPLE 1: States: "The Child shall enjoy all the rights set forth in this declaration. Every Child without any exception whatsoever shall be entitled to these rights, without distinction or discrimination on account of race, colour, Sex, language, religion, or other opinion, national or social origin, Property, birth, or other status, whether of himself or of his family".

Perhaps, a fundamental notice should be taken in the stated word in Principle 7 which states: "The child. Shall be given an education which will promote his general culture and enable him on a basis of equal opportunity to develop his abilities, his individual judgment and his sense of moral and social responsibilities and to become a useful member of society.

Principle 9, thereby states, "The Child shall not be admitted to employment, before an appropriate minimum age; he shall in no case be caused or permitted to engage in any occupation or employment which would prejudice his health or education, or interference with his physical, mental or moral development'.

Based on these principles of our children's well-being our company has developed an Educational Development Master Plan, (EDMPI). This program is the product of a research by our company to understand better the problems of students who are in their mid-year in high school (secondary school). We discovered that many students leave school in a confused state, as to what appropriate career to be studies or practiced. So, they rely on parents, counselors or mere peer-pressure to satisfy their yearnings. Based on professional experiences, records and most importantly on the job practice and interaction, we innovated the strategic plan to help solve these problems. We are sure to predict at a 95% accuracy where a student should be or at least in a related field. The child shall be guided by appointed officers to fill the necessary forms.

PURPOSE AND PROJECTIONS OF THE EDMP.

- We hope to bring about positive change in the state and national education sector by helping a child to discover his potentials, more especially as a teenager.
- We hope to eradicate the era of misplaced priorities. Only by special causes, should an engineer find himself in a contrary field of operation.
- By our discovery, the early awareness of one's prospect career open to him, confidence is developed, innovations are introduced, cheating is discouraged and a host of students negative excesses are curtailed.
- Focus are developed and the drift towards unruly activities and behaviour such as stealing, robbery and terrorism is eradicated because he wants to fulfill his obligations to the community.
- By this records kept, it is easier to trace terrorists in any city in the future which would open our eyes better to other reasons of such behaviours, as parents would be held responsible for their children's behaviour.
- We see a world where the SCAP sheets are attached to other relevant records during a job application as part of a Curriculum Vitae.
- In order to curtail the effects of man's inhumane treatment on earth, we hope to recommend the study of geography (recommended subject), and agricultural science (science students), creative sciences as well as education, history (recommended course department).
- Needless to say, there is a massive employment apparatus for the state if you support us. This exists in the departments of analysis (by marking); record-keeping, advisory unit, management operations, transfer

services and information technology, most especially in infrastructural development and training for our special regulatory measures.

- We have our hands on the job experience with an overwhelming response as schools have been analysed and success recorded.
- It is also easier to monitor progress of the participants on an individual and state/ society rating.
- Lastly, the spirit of self reliance, entrepreneurship and less dependence on the government to provide jobs is fostered.

It is required that the program is approved by the state Government compulsorily for all Secondary Schools eligible participants. We are pen to further negotiations.

Thanks.
Peter Matthews-Akukalia Ch.s
Author
President/CEO.

THE PRODUCT
TITLE: Students Career Analysis Program (SCAP)

FACILITATORS: Utmost Educational/ Consult Services
(A Subsidiary of Utmost Peak Resources)

PURPOSE: To give students a career direction and enhance confidence in career selection.

GUIDING PRINCIPLES OF SCAP
The guiding principles of the research discovery and creation of "SCAP" are categorized into three groups:

- Elites Intellectual Philosophies
- The United Nations Child Rights Acts (As Quoted)
- The Company's Contributive Developmental policies to enhance educational standards.

1. ELITES INTELLECTUAL PHILOSOPHIES

- The common curse of mankind is folly and ignorance William Shakespeare
- The foundations of knowledge must be laid by reading. General principles must be had from books, which however must be brought to the test of real life-Johnson.
- Who reads incessantly and to his reading brings not a spirit to judgment equal or superior, uncertain and unsettled still remains-Milton.
- Fields and trees teach me nothing, but the people in a city do-Euripides
- A true-bred merchant is the best gentleman in the nation-Robinson Crusoe
- The foundation of every state is the education of its youths-Diognes
- Never educate a child to be a gentleman or a lady only, but to a man; a woman. Herbert Spencer.
- Our studies pass out of sight into character-Anonymous
- Education is too important to be left solely to the sole educators-Journalist
- Never educate one truncatedly and one-sided-E.M.P. Edeh

- He who opens a school door closes a prison's door- Mahatma Ghandi
- The career open to talents-that was principle Napoleon

2. DECLARATION OF THE RIGHTS OF THE CHILD

   On November 20, 1959, the United Nations General Assembly unanimously approved and proclaimed the Declaration of rights of the Child, which lays out the rights and liberations that all children without exception should enjoy.

   Many of these rights and liberations are already mentioned in the Universal Declaration of Human Rights adopted by the General Assembly in 1948. However, they recognized that the special needs of children justified separate declarations.

   In 1998, the United Nations General Assembly approved the convention of the Right of the Child, which grew out to the declaration of the Rights of the Child. The 1998 convention was been ratified by the majority of member nations, Nigeria is a signatory to this charter.

THE GENERAL ASSEMBLY

PROCLAIMS this Declaration of the Rights of Child to the end that he may have a happy childhood and enjoy for his own good and for the good of society the rights and freedoms herein set forth, and called upon parents, upon men and women as individuals and upon voluntary organizations, local authorities and national governments to recognize these rights and strive for their observance by legislative and other measures progressively taken in accordance with the following principles:

PRINCIPLE 9:7(a)

The Child is entitled to receive education… He shall be given an education which will promote his general culture and enable him on a basis of equal opportunity to develop his abilities, his individual judgment and his sense of moral and social responsibilities and to become a useful member of society.

3. UTMOSTEDUCATIONAL DEVELOPMENT POLICIES ON SCAP

- To encourage the compulsory participation of concerned students in SCAP.

- To encourage reading culture-which informs the reason for the free copy of the company's published Annual Year Book to all participants.
- To encourage the study of Geography-this is in order to sensitize the younger generations on the effects of man's activities in respect to world climate changes, soil science in respect to the recent world food crisis and agriculture; and better understanding of Globalization in business and occupations; and as well as the appreciation of various cultures of different people.
- To resuscitate the study of History this to inculcate the attitudes of patriotism, appreciation of the human person and diversity, and the development of healthy personal confidence in building upon the heroic deeds achieved by great people, and striving for excellence without bias or fear.
- To encourage the study of our innovative course on "Creative sciences" to help them appreciate and develop their potentials.

The study of Geography also cuts across every cadre of studies i.e.

- Sciences Geo-physics, Engineering etc.
- Arts/ Humanities- International law and Diplomacy; Linguistics; Cultural Anthropology etc.
- Commercial (Administration & Social Sciences)- geography and Environmental Management: Economics and Geography; Development Studies. Etc.
- There is also no prejudice of any Child's abilities since he makes his own future by his own decisions as outlined on paper.
- The discovery of each child's own potentials both in the Academic based and vocational fields of practice.

NATURE SCAP

The Students Career Analysis Program was designed in consonance with International standards. It comprises Six forms known as Sheets, out of which Four must be filled by the student independently but guided by an approved officer. The two extra sheets are results. They are detailed below:

SHEET Aa1: PERSONALITY REPORT

This sheet describes the name, age, gender, personality traits (strength and weaknesses) of each participant; health, date of analysis.

SHEET Aa2: DISABILITY PROFILE

It is practically understandable that a blind man cannot drive a car. This page comprises certain general illness which may or usually confront a child. The illnesses stated are thus: Eyesight, Hearing, Skin complaints, Epilepsy, Travel Sickness, Dizziness at heights, and lack of physical strength. The extent of these illnesses as suffered by the student is also defined by the student. Three boxes are placed bearing the terms: very much, little or sometimes and Nil, as well as observed by the trained official.

SHEET Ba1: INTERESTS AND ACTIVITIES DATA BASE

Every hobby often practiced by a person will more often lead to a field of endeavour or career. This page contains the recognized hobbies known as activities as well as code numbers which guide to their known fields. Each student/ participant is expected to select activities which tickles his interest or preference. This is done by ticking.

SHEET Ba2: SUBJECT CHOICE CHART

There are four boxes in this chart stated in line with the subjects done at each level. The boxes include (i) Tick those you enjoy (ii) Tick those you are good at (iii) Reasons (iv) Name of Subject Teacher. Those selected grant a basic guide to the selection of the academic based courses in line with other information. Schools with little or not enough teachers may be discovered from this list.

RESULT SHEET

The result sheets are divided into two:

i) Sheet Ca1: This contains the name of student, name and address of school and a section graph. The section graph is plotted to fined to find the point of highest interest of a student.

ii) Sheet Ca2: This contains the recommendation section which include spaces for the recommendation of academic course as well as vocations which may be fitting for the student according to the section graph of sheet Ca1.

The page also contains the fitted departments of sciences, arts and commercial which the company recommends after the analysis.

MASTER SHEET

The master sheet was designed by the company to facilitate record-keeping. It is a sheet of two sides compressed to contain all the records of a student. It also contains the information of who last entered an information, date of analysis, date of entry, year book information. This ensures accountability and responsibility. When the results are released to students, master sheets are given to Principals of schools while copies are kept by the company. It is hoped that the company shall send resident Counselors (R.C) to schools for a period of about Seven (7 days) days to help students understand their results as well as guide them better, or provide a special Dictionary of careers (Local & foreign) for their personal use.

LEVELS/ CATEGORIES OF SCAP

The SCAP was developed to accommodate three levels of people:

(A) Junior SCAP: Specifically meant for the JSS 3 only. It is advisable that schools and states participating for the first time make allowance for their SSS 1 to participate along. It is an Annual Program and all students ought to consider it the full essence of their entire education.

(B) Senior SCAP: Meant for schools participating the first time only. It is designed for the Ss2 and Ss3 participation. The difference here is that students tick their already chosen subjects/ departments of which the academic based course and vocations are determined. All other factors remain the same.

C) Adult SCAP: Also known as the Human Resources / Personality Predictability Perspective (HR/ P3) Analysis. Designed for staff development, it aids better understanding of an builds a profitable relationship between employer and employee. It also helps the management to define proper roles for staff in relation to their interest and other factors discovered. This contains all factors with an advanced questionnaire. It is designed for adults, graduates and undergraduates who sincerely wish to know their own abilities.

BENEFITS OF SCAP

The benefit and importance of SCAP are classified into four parts and categorized into three: (i) Short Term (ii) Medium Term (iii) Long Term

1. The Student / Participant

Short Term Benefit.

- Freedom to make determined and informed choices
- Early discovery of self and abilities
- Confidence in choosing a career
- Opportunities to ask questions and on fields conduct research of interest.

2. Medium Term Benefits

- Tends to consider studies more seriously
- The aptitude for design and research is enhanced

3. Long Term Benefits

- Spends less time in determining vocation or further studies
- Learns early independence
- Learns the spirit of responsibility
- Less dependence on government for job
- Self-employment and eradication of poverty.

2. The Parent

- The parents better understand their Children
- There is hope determined as to plans for their Children
- There is less blame on government or people for misfortunes or a child's misbehaviour.
- They see value for their investment.

3. The School

- There is less work-stress for the school to do
- Assistance of resident counselors from company to see to welfare of students.
- Students easily appreciate the education and efforts offered by school authority.

4. The State

- Better educational rating by the United Nations
- Contributive to our Millennium Development Goals (MDGS)
- Mass employment and training of staff
- Parents can now be held responsible for their children's behaviour
- Crime is reduced drastically
- Poverty is economically reduced to a barest minimum.
- Policy making can now be more effective when people are willing to strive for the better, ideals of hard work, integrity, faith and transparency.
- Diversification of the economy
- General empowerment of people.

EXTRA FUNCTIONS OF THE COMPANY

1. When contracted, the company shall perform advisory services to the state on the development of certain neglected vocations and centers. This shall lead to the diversification of the economy.
2. The company when contracted shall help to promote these skills acquisition centers.
3. When contracted the company research department shall provide incentives for students who are inventive, creative or innovative in their intellectual and vocational researches.
4. The company shall contribute its best in term of efforts and further research to see that our youths do their nation proud in their vocation and career pursuits.

PREVIOUS RECORDS

As an enterprise, the company has been able to carry out this program extensively with an overwhelming acceptance in the Private Sector. The state can also benefit from this project.

FUTURE PROJECTIONS

- We see a nation whereby the SCAP becomes a part of an applicant's Curriculum vitae (C.V.) in order to avoid misplaced priorities.
- We see a nation where youths are intellectually challenged to confront their future in confidence.

- We see a more vibrant state/ nation in accordance with the ideals of the United Nations Millennium Development Goals.

COST IMPLICATIONS

- Subject to state and company negotiations, and company decisions for the time being in force.
  Everyone is important! According to William Shakespeare's Julius Ceasar, "Who is here that will not want his child to be give the best education?"

UTMOST EDUCATIONAL/ CONSULT SERVICES

Subsidiaries of utmost peak resources.
20,Moriamo Adesina Street, Off Asalu Street, Asalu B/stop,
Abaranje Road, Ikotun-Lagos
Tel: 08022045323, 07084289130, 07039284942
E-mail utmostpr@yahoo.com

STUDENTS CAREER ANALYSIS PROGRAM (SCAP)

SHEET Aa1

Date: ______________________

PERSONALITY REPORT ON:

1. NAME: ______________________________________
   (Surname) *(Other Names)*
2. AGE ____________________ GENDER: Male Female
3. HEALTH: Tick as appropriate Fit Disabled Sick
4. ACTIVITIES THAT I DO IN MY SPARE TIME
   i. ______________________
   ii. ______________________
   iii. ______________________
5. MY GOOD POINTS ARE: (choose from the Options below)
   (1) (2)
   i. ______________________
   ii. ______________________
   iii. ______________________
6. MY WEAK POINTS ARE: (choose from the Options below)
   i. ______________________
   ii. ______________________
   iii. ______________________

(1)	(2)
A. Moody	**A. Slow thinking ability**
B. Selfishness	**B. Poor Writing**
C. Laziness	**C. Disobey Instruction**
D. Unfriendliness	
E. Clumsy	
F. Rude	
G. Silly	
H. Messy in appearance dress, attitude & work	

SHEET Aa2
DISABILITY PROFILE

Which of these disabilities do you suffer from?
Tick as appropriate

1. **Eyesight**	**Very Much**	**Little or Sometime**	**Nil**
2. **HEARING**	**Very Much**	**Little or Sometime**	**Nil**
3. **SKIN COMPLAINTS e.g DERMATITIS, ECZEMA**	**Very Much**	**Little or Sometime**	**Nil**
4. **ASTHMA**	**Very Much**	**Little or Sometime**	**Nil**
5. **TRAVEL SICKNESS**	**Very Much**	**Little or Sometime**	**Nil**
6. **TRAVEL SICKNESS**	**Very Much**	**Little or Sometime**	**Nil**
7. **DIZZINESS AT HEIGHTS**	**Very Much**	**Little or Sometime**	**Nil**
8. **LACK OF PHYSICAL STRENGTH**	**Very Much**	**Little or Sometime**	**Nil**

Being realistic now will save you disappointment later.

SHEET Ba1
INTERESTS AND ACTIVITIES DATABASE

What do we mean by activities?
Not Just School subjects but things you enjoy doing
In or out of school.

Tick any activities you enjoy	Tick here	Office use only careers- Basic codes			
1. Wood Work or metal work?		2	2	5	7
2. Typing/ Computer works		4	6	8	
3. Gardening/ Flower Arranging?		2	3	5	7
4. Repairing Mechanical or Electrical items?		2	3	5	7
5. Playing Sports?		2	3	5	7
6. Pottery, Drawing or Painting		5			
7. Cooking		1	2	3	
8. Sewing, embroidery, tapestry, e.t.c		3	5		
9. Music or Drama		1	5		
10. Helping to run a club?		1	4	6	8
11. Community service e.g. helping old people		1			
12. Fishing, walking, sailing, etc.		3	7		
13. Astronomy, Photography?		2	5	6	
14. Writing or reading stories or poem?		4	5	8	
15. Belonging to an Association e.g. scouts guides, St Johns Ambulance.		1	8		
16. Do it yourself, e.g. Painting/ Decorating?		3	5	7	
17. Collecting things and indexing them		2	4	8	
18. Playing indoor games, e.g. Chess?		6	8		

SHEET Ba2
SUBJECTS CHOICE CHART

SUBJECT	Tick Those you enjoy	Tick Those You Are Good At	Reasons	Name of Teacher
ENGLISH LANGUAGE				
MATHEMATICS				
HOME ECONOMICS				
SOCIAL STUDIES				
PHYSICAL / HEALTH ED.				
YORUBA				
IGBO				
HAUSA				
LITERATURE IN ENG.				
BUSINESS STUDIES				
LAGOS STUDIES				
COMPUTER STUDIES				
INTRODUCTORY TECHNOLOGY				
INTERGRATED SCIENCE				
FINE ART				
AGRICULTURAL SCI.				
SEXUALITY EDUCATION				
RELIGION EDUCATION				

SHEET Ba2
SUBJECTS CHOICE CHART

SUBJECT	Tick Those you enjoy	Tick Those You Are Good At	Tick those you feel you may want to study
ENGLISH LANGUAGE MATHEMATICS RELIGION EDUCATION PHYSICAL EDUCATION HISTORY GEOGRAPHY ART NEEDLE WORK LANGUAGE STUDY WOOD WORK METAL WORK TECHNICAL DRAWING DRAWING COOKERY PHYSICS BIOLOGY ECONOMICS CHEMISTRY AGRICULTURAL SCIENCE FURTHER MATHEMATICS FINANCIAL ACCOUNTING COMMERCE GOVERNMENT LIT. IN ENGLISH			

SHEET Ba3

PERSONAL ATTITUDE CHART (HR / P3 ANALYSIS only)

Put a ring around Nos 1,2,3,4, or 5 to indicate how well you could cope with the questions posed

How good are you at coping with	Not at all	Not Very	Not Sure	Quit Good	Very Good
SECTION A					
1. Your own problem?	1	2	3	4	5
2. Other People's Problem?	1	2	3	4	5
3. Not being appreciated?	1	2	3	4	5
4. People in a Bad mood?	1	2	3	4	5
5. Other wanting their own way?	1	2	3	4	5
6. Being friendly to all sorts of people?	1	2	3	4	5
SECTION TOTAL SCORE =					
SECTION B					
7. Detailed accurate work?	1	2	3	4	5
8. Memorizing facts?	1	2	3	4	5
9. Writing Reports?	1	2	3	4	5
10. Telephones or Office Machinery?	1	2	3	4	5
11. Situations needing instant decisions?	1	2	3	4	5
12. Work under Pressure?	1	2	3	4	5
SECTION TOTAL SCORE =					
SECTION C					
13. Working under difficult conditions?	1	2	3	4	5
14. Long Hours?	1	2	3	4	5
15. Routine Repetitive Work?	1	2	3	4	5
16. Taking instructions from others?	1	2	3	4	5
17. Physical Exertion?	1	2	3	4	5
18. Little Chance of Promotion?	1	2	3	4	5
SECTION TOTAL SCORE =					

UTMOST EDUCATIONAL/ CONSULT SERVICES

Subsidiaries of utmost peak resources.
20,Moriamo Adesina Street, Off Asalu Street, Asalu B/stop,
Abaranje Road, Ikotun-Lagos
Tel: 08022045323, 07084289130, 07039284942
E-mail utmostpr@yahoo.com

SHEET Aa1

STUDENT'S CAREER ANALYSIS

RESULTS SHEET

NAME OF STUDENT: ____________________

NAME OF SCHOOL: ____________________

ADDRESS OF SCHOOL: ____________________

AGE ______________ GENDER: Male   Female

STUDENT'S INTERESTS & OCCUPATIONS GRAPH

A								
B								
C								
D								
E								
F								
G								
	1	2	3	4	5	6	7	8

CARING OCCUPATIONS
SCIENTIFIC WORK
SKILL WITH HANDS
JOBS USING ENGLISH
CREATIVE WORK
FIGURE WORK
OPEN AIR /PACTICAL JOBS
OFFICE BASED/ CLERICAL WORK

* This section-graph indicates the Student's interest based on collections from database.

Occupation	Points
1. Caring	
2. Scientific Work	
3. Skill with Hands	
4. Jobs using English	
5. Creative Work	
6. Figure Work	
7. Open Air/ Practical Jobs	
8. Office-Based/ Clerical Work	

SHEET Ca2
RANGE OF CAREERS OPEN TO STUDENT

1. ______________________________
2. ______________________________
3. ______________________________ Academic-based
4. ______________________________
5. ______________________________
6. ______________________________
7. ______________________________
8. ______________________________ Vocational
9. ______________________________
10. ______________________________

RECOMMENDED SUBJECTS AT SENIOR SECONDARY LEVEL (Tick Only)

1. MATHEMATICS	7. Physics	Government	Commerce
2. ENGLISH LANGUAGE	8. Chemistry	Literature	F. Accounts
3. BIOLOGY	9. Religion Studies	Agric science	
4. ECONOMICS	10. Further Maths	(Optional)	
5. GEOGRAPHY	11. History	(optional)	
6. LANGUAGES			

RECOMMENDED SUBJECTS DEPARTMENT

1. SCIENCES
2. ARTS
3. COMMERCIAL

For Further enquiries, please visit in person to the above address with an enclosed application fee of ______________________________

____

Authorized Sign, Stamp & Date

UTMOST EDUCATIONAL/ CONSULT SERVICES

SUBSIDIARIES OF: UTMOST PEAK RESOURCES ®

20, Moriamo Adesina Str. Off Asalu B/Stop,
Abaranje Road, Ikotun, Lagos. Nigeria
Tel: 08022045323, 07084289130

STUDENTS CAREER ANALYSIS PROGRAM (SACP) RESULTS

MASTER SHEETS PM.1

PARTICIPANT /STUDENT DETAILS

Name of student ______________________________
Date of Birth: (DD)________________Age ______________
Gender (TICK) MALE: Female
Health Status (TICK): FIT Disabled Sick
Nature of SCAP (TICK) JUNIOR Senior Adult

Nationality:__________________ZIP CODE:______________
Local Govt. (NIGERIAN ONLY):______________________
State of Origin (NIGERIAN ONLY):____________________
Name of School/ Organisation:_______________________
Address of Schools/ Organisation:____________________
Date of Analysis:________________Date of Return:__________
State Disabilities (If Any) & Extent:___________________
Student's Interests (NOS, ONLY):_____________________
Subjects Enjoyed:______________________________
____
Subjects Good At: ______________________________
____
Hobbies/ Activities of interest:
I. ____
ii. ____
iii. ____

CERTIFIED COPY

RESULTS

INTERESTS DATA BASE (SECTION-GRAPH)

OCCUPATION

A Caring
B Scientific Work
C Skill With Hands
D Jobs Using English
E Creative Work
F Figure Work
G Open Air/Practical Jobs
H Office Based/ Clerical Work

1 2 3 4 5 6 7 8

OCCUPATION BLOCK

OCCUPATION POINTS

1. Caring
2. Scientific Work
3. Skill With Hands
4. Jobs Using English
5. Creative Work
6. Figure Work
7. Open Air/Practical Jobs
8. Office based/ Clerical work

***RECOMMENDED SUBJECTS AT SENIOR SECONDARY LEVEL (TICK ONLY)**

1. ENGLISH LANGUAGE
2. MATHEMATICS
3. BIOLOGY
4. ECONOMICS
5. **GEOGRAPHY
6. LANGUAGE (AS APPLICABLE)
7. PHYSICS GOVERNMENT COMMERCE
8. CHEMISTRY LIT IN-ENG. F. ACCOUNTS
9. RELIGION STUDIES AGRIC. SCI
10. FURTHER MATHS (OPTIONAL)

11. History (OPTIONAL)

* Recommended subjects and dept. are NOT suitable for adult SCAP (HR/P3 Analysis).

* Special recommendation.

RECOMMENDED SUBJECTS DEPARTMENT (TICK ONLY)

1. SCIENCES 2. ARTS 3. COMMERCIAL

DATE OF ENTRY: DAY ________ MONTH ______ YEAR __________

SESSION ____________________

NAME OF OFFICER & SIGN: ______________________________

STATE: ____________________COUNTRY __________________

YEAR BOOK GIVEN? (TICK) YES NO CONDITIONAL

NATURE OF BOOK (TICK)

1. PLAY
2. PROSE
3. POETRY
4. CERT OF PARTICIPATION C.O.P
5. DICTIONARY OF CAREERS
6. PROPOPLAYI
7. NIDRAPOE
8. PRINCIPLES BOOK
9. SUBJECTS

Certified Copy

Authorized Stamp
Sign & Date

RESULTS

ARTS/ HUMANITIES /LAW

RECOMMENDED CAREERS

1. LAW

1.*____________________ ACADEMIC-
2.*____________________ BASED
3.____________________
4.*____________________ VOCATIONAL
5.____________________
NOT RECOMMENDED? TICK HERE

2. EDUCATION

1.____________________ ACADEMIC-
2. ____________________ BASED
3.____________________
4.____________________ VOCATIONAL
5.____________________
NOT RECOMMENDED? TICK HERE

3. ARTS

1.*____________________ ACADEMIC-
2. *____________________ BASED
3.____________________
4.*____________________
5.____________________ VOCATIONAL
NOT RECOMMENDED? TICK HERE

4. LANGUAGES

1.*____________________ ACADEMIC-
2. *____________________ BASED
3.____________________
4.*____________________
5.____________________ VOCATIONAL
NOT RECOMMENDED? TICK HERE

Any Alteration renders this document invalid
*Foreign Studies and Vocations only. Scholarships into foreign universities and colleges are available.
Please consult the school authority.

Certified copy

Authorized
Sign & Date

RESULTS

COMMERCIAL

RECOMMENDED CAREERS

1. ADMINISTRATION

1.*____________________	ACADEMIC-
2.*____________________	BASED
3.____________________	
4.*____________________	VOCATIONAL
5.____________________	

NOT RECOMMENDED? TICK HERE

2. SOCIAL AND MANAGEMENT STUDIES

1.*____________________	ACADEMIC-
2.*____________________	BASED
3.____________________	
4.*____________________	VOCATIONAL
5.____________________	

NOT RECOMMENDED? TICK HERE

3.EDUCATION

1.*____________________	ACADEMIC-
2.*____________________	BASED
3.____________________	
4.*____________________	VOCATIONAL
5.____________________	

NOT RECOMMENDED? TICK HERE

4. AGRIICULTURE

1.*____________________	ACADEMIC-
2.*____________________	BASED
3.____________________	
4.*____________________	VOCATIONAL
5.____________________	

NOT RECOMMENDED? TICK HERE

Any Alteration renders this document invalid
*Foreign Studies and Vocations only. Scholarships into foreign universities and colleges are available.
Please consult the school authority.

Certified copy

Authorized
Sign & Date

RESULTS

SCIENCES

RECOMMENDED CAREERS

1. MEDICAL AND PHARMACEUTICAL SCIENCES

1.*________________________________ ACADEMIC-
2.*________________________________ BASED
3.________________________________
4.*________________________________ VOCATIONAL
5.________________________________
NOT RECOMMENDED? TICK HERE

2. ENGINEERING / TECHNOLOGY/ ENVIRONMENTAL SCIENCES

1.*________________________________ ACADEMIC-
2.*________________________________ BASED
3.________________________________
4.*________________________________ VOCATIONAL
5.________________________________
NOT RECOMMENDED? TICK HERE

3. AGRICULTURE /EARTH SCIENCES

1.*________________________________ ACADEMIC-
2.*________________________________ BASED
3.________________________________
4.*________________________________ VOCATIONAL
5.________________________________
NOT RECOMMENDED? TICK HERE

4. EDUCATION

1.*________________________________ ACADEMIC-
2.*________________________________ BASED
3.________________________________
4.*________________________________ VOCATIONAL
5.________________________________
NOT RECOMMENDED? TICK HERE

5. GENERAL SCIENCES

1.*____________________	ACADEMIC-
2.*____________________	BASED
3.____________________	
4.*____________________	VOCATIONAL
5.____________________	

NOT RECOMMENDED? TICK HERE

Any Alteration renders this document invalid
*Foreign Studies and Vocations only. Scholarships into foreign universities and colleges are available.
Please consult the school authority.

Certified copy

Authorized
Sign & Date

BRIEF INTERPRETATIVE FINDINGS (BIFS)
(HR/P3 ANALYSIS ADULTS ONLY)
PRIVATE S' OFFICIAL USE

PERSONALITY PREDICTABILITY PERSPECTIVE (P3)
LEVELS; TICK ONLY

1. **NATURE OF GRAPH:** HIGHLY SIMPLE SIMPLE
HIGHLY COMPLEX COMPLEX
2. **RESOURCEFULNESS:** AVERAGE MEDIUM HIGH
3. **BRILLIANCE/ INTELLIGENCE RATING (BIRA):** AVERAGE MEDIUM HIGH
4. **DECISION MAKING ABILITY (DEMA):** AVERAGE MEDIUM HIGH
5. **PRACTICALITY RATIO (PART):**
AVERAGE MEDIUM HIGH
6. **SOCIALS:** AVERAGE MEDIUM HIGH
7. GENERAL INCLUINATION OF LITERARY
TECHNICIAL LI-TECH
8. LEVEL OF ENDO WMENT (LOE):
A.P. = 100%

PEAK OF ASPIRATION ON (POA)
=1,000% IE
=____________%IE

DESCRIPTION: AVERAGE MEDIUM HIGH

Certified copy

____

Authorized stamp
Sign & Date

CHAPTER 20

WHAT IS EDUCATION?

AIMS/ OBJECTIVES:

1. To improve on the present system of education through applying the creative process.
2. To view education as an integral part of the creative process.
3. To revolutions the educational process with the inclusion of the special creative knowledge.
4. To re-define education by understanding the discovered creative laws.
5. To empower the educational process to be more useful and applicable through skills development.
6. To discuss the responsibilities, problems, likely solutions of a poor educational system through the scientific application of creative analysis available to the government and citizenry.
7. To prove the difference between functional and non-functional education in every society through results.
8. Functional education must be evaluated by the results and inputs of the educated.
9. To teach the different abilities inherent in every person academically.
10. To help recognize the difference between brilliance and intelligence.

WHAT IS EDUCATION?

I shall begin this chapter by defining the word 'education'; from the conventional views, after which I do hope to do justice, on paper, to this topic. Remember, I have applied my common sense to this issue with the sole aim of dissecting and masticating the piece in order to help us redefine by evaluation the present ground on which we stand. This, I have done by applying my Sense of institution, perception and observation. The standard of measurable excellence lies not in the results but in such process that if a popular method prescribed and proscribed were not effective, in a sector of national means, an institution of human learning, a school, a classroom, or, perhaps a vocational or skill acquisition centre; then better creations and innovations should be established.

According to the American Heritage Dictionary, it defines and explains Education as; "1 The act or process of educating or being educated. 2. The knowledge or skill obtained. 3. The field of study concerned with teaching and learning", for emphasis, the word" educate" is explained thus. "1. To provide especially with formal knowledge or training 2. To provide with information. 3. To bring or known as an educator."

The British Collins Gem English Dictionary says that, by implication, the word education came from educate and this means "teach, provide schooling for" A philosopher once said, " The surest way into truth is by perception, by intution, by reasoning to a certain point, then by taking a mortal leap, and by intuition attaining the truth". Perhaps, we shall need to examine carefully certain principles of education as made or uttered by the masters and see, if perhaps, we way gain certain insights into their mastery.

1. The man who thinks only of himself is hopelessly uneducated.-Dr. Nicholas M. Butler. This implies that one with a proper education must seek to share, either his knowledge or related issues so as to benefit others. Confidence, not conceit is the guiding principle.
2. Education is too important to be left solely to the sole educators.-Journalist.
   Here, everyone has a share, a duty in the matters related to education.
3. Utter truth is essential, and to get that truth may take a lot of searching and long hours.- Ellen G. White.
   Obtaining a thorough knowledge is a worthy goal, but one sure does need patience to achieve this.
4. The foundation of every state is the education of its youths.- Diognes, It is the first duty of every country or nation to settle first the issues of education, or else, the mighty structure being placed on it will eventually collapse one day.

5. Education is discipline for the adventure of life; research is intellectual adventure; and the universities should be homes of adventure shared in common by young and old. Anonymous
6. I believe the object of education is the freedom of mind, which can only be achieved through the path of freedom-though freedom has its risks and responsibility as life itself has. _______ Tagore.
7. The secret of Education lies in respecting the pupil Emerson.
8. The pen is mightier than the sword Shakespeare.
9. Our studies pass out of sight into character.
10. Education makes a people easy to lead, but difficult to drive; easy to govern, but impossible to enslave Henry Brougham B. Ham.

I shall pen down my thoughts so piled, till some great truth is exposed, and the nations echo round, shaken to the roots.

Firstly, I shall define a definition as" "A statement conveying fundamental character; a statement of the meaning of a word, phrase, or term, as in a dictionary entry; the act of making clear and distinct; a determination of outline, extent, or limits", according to the American Heritage Dictionary. From this, we may deduce two kinds of a definition as; definitive and indefinitive.

1. **Definitive Definition:** Also, known as conclusive it is based on proved facts, by principles and practice. It is sure based on intelligent scrutiny and yields itself as a better result each time. It is not narrow in structure of its existing patterns.

2. INDEFINITIVE DEFINITION:

Also referred to as inconclusive. It is often characterized on blamelessness or narrow mindedness. Though subject to further scrutiny, its under lying foundation provides no tangible base for further investigations.

Based on this background, I shall endeavor to masticate the spicy delicacy presently laid before me. Education is the process of imparting knowledge to another (in definitive) why so? The sole aim of education is to train the mind of a living entity (such as a person or even animal; consider, the instance, where by a dog is trained; to detect at airports or borders certain banned chemical substances such as cocaine, heroine etc; or, perhaps made to behave in certain manner at certain times) to learn better about himself and his existing environment: the person who imparts knowledge, not considering the circumstances, whether in the shop, church, mosque, school, road etc; may be called an Educator as long as he is able to communicate within appropriate means. The one who is willingly learning, I shall describe as the Acceptor or learner. For one to be educated, he must show willingness. He must be willing to accept the imparted knowledge or training morally and psychologically. Simply put, the act of willingness creates <u>interest</u> in the subject matter being discussed.

Therefore, I shall define education definitively as; the process of impartation of knowledge to an interested acceptor". If this is not so, the education is a noise, the educator sounds senseless and becomes a bore.

Further, Education would be said to involve two parties whose interests, reciprocates each other. In other words, the one who is educating and ultimately teaching should be happy to teach and as well the one receiving or accepting should also be happy to do same.

ANALYSIS OF EDUCATION

Analytically education may be expressed as;
Education= EDUCATOR interest ACCEPTOR

But knowledge is obtained at a place and within a specific time, Pmakumatically, it may be expressed as;

EDUCATION= Education (ET) +ACCEPTOR (AT)
Time (T) X Place (P)

Or: E= ET + AT
KTP ; Where K is a sign for constancy, because time and place define our environment which is constant.

This is what I shall refer to as: P.M. Aku's formula of educational Analysis.

PURPOSES OF EDUCATION

Every thing created, innovated, produced or ever made has its purpose. A thing without a purpose is useless and not fit for anything useful. I shall, in my candid discovery propose, the two major purpose of Education. These are Learning and Adaptability.

LEARNING:

What is learning?

The British Collins Gem Dictionary describes it as, "gain skill or knowledge by study, practice, or teaching; memorize (something); find out or discover (learn); while knowledge got by study (learning)".

The American heritage Dictionary defines "learn" as; "To gain knowledge, comprehension, or mastery of through experience or study; to memorize; to become informed; or perhaps, to discover synonymously", while "Learning" is defined as "Acquired knowledge or skill", Jerry Carroll, the president, Airman Flight School, Oklahoma, U.S.A; in his letter to me in 2005 stated "Learning is defined as a change in behaviour based on experience", while "learning may be defined as the process by which an activity originates, or is changed, through reacting to an encountered situation. Thus the type and quality of the experience is important in structuring a learning situation". The first requirement is to obtain interest and attention; trainees must be convinced of the need to learn the skill and progress of the trainee will depend on;

a) An appreciation of the nature of the activity involved.
b) The complexity of the skill
c) The quality of the performance the trainee has as a model
d) The trainee's natural endowment (aptitude).

The acquisition of a skill is characterized by a progressive change in performance; the final proficiency is not merely the performance at the beginning carried out more rapidly; it is a different performance. In addition to the time decrease, there should be also;

a) A reduction in muscular effort.
b) An elimination of surplus movements.
c) An awareness of progress

d) Greater self-confidence
e) Greater precision.

The trainee should develop, as a result of this process the ability to anticipate and plan ahead-performance becomes smoother and speed increases without any appearance of hurry. Learning a skill is not a mechanical repetition of actions, it is an educative process involving understanding. As a result, the trainee is psychologically adjusted to the job, as well as functionally proficient" (GEORGE ALLEN & UNWIN-AUTHORS of A First course in business organization). One who teaches must first learn; so also the one who is being taught must also learn. I therefore define learning as; the process of absorbing by reception and observation certain necessary information and experiences within an environment with a view to achieving a certain necessary purpose.

Learning is not complete without a realization that one has received, with his consent; a certain knowledge; something tickling his fancy, or baffling his thinking. Proved. That is why an intelligent person karns by pereception (That extra sensory ability), Observation and foresight. The intelligent person may not necessarily be brilliant, or excelled at school work, Intelligence deals primarily with am ability to make adept decisions, changing and behaving appropriately to a new encountered situation; whole brilliance conforms itself to passing curtain required tests of proficiency. A brilliant person (Pupil, or student) may well be good at school work, but, perhaps, not necessarily able in solving simple problems he faces in every day practical life; or do you not read of the "bookful block head" in the following verse:

> "Deep versed in books and shallow in himself, the book full blockhead ignorantly read, with loads of leaned number in his head" __ Alexander Pope.
>
> Perhaps we might be able to distinguish the intelligent from a mere brilliant in the verses;

"who reads
Incessantly and to his reading brings not a spirit or judgment equal or superior......... uncertain and unsealed still remains". Mitton:

Learning by reading may be better understand when we see this instruction verses; "Read not to contradiction and confute; nor to believe and take for granted all, nor to find talk and discourse; but to weigh and consider; Francis Bacon

Then Johnson complement on knowledge "The foundations of knowledge must be laid by reading. General principles must be had from books, which however must be brought to the test of real life."

"Books are not absolutely dead things, but do contain a potency of life in them, to be as active as that soul whose progeny they are. Nilton.

While Francis Bacon note's in practical terms: "History makes men wise; poetry, witty; the mathematics, subtle; natural philosophy, deep; moral, grave; Logic and Rhetoric, able to contend".

Then that respected Ambassador of education caps it with a solemn speech;
"I value education in the different Sciences; our children cannot have too much of chemistry and Physics." Mahatma Ghand: of curse, he who is tired of learning is tired of life for life is full of lesson and is a lesson in itself. Based upon these profound truths and others, I hereby propound the 16 laws of learning.

P.M. AKU'S LAWS OF LEARNING;

1st Law States:
Learning is never complete without a basic interest of the acceptor; this interest is obtained by consent; and the willingness of the educator.

2nd Law state;
Learning is achieved best when the Acceptor's interest rise to the extent of further enquiries, or asking questions.

3rd Law state:
Learning is a barbaric act when it is not followed by the process of understanding; for this strengthens the memory.

4th Law states:
Learning is independent of situations, or environmental factors as long as it is favoured by it, this includes concentration, quetress, and realization of the need to learn. With reference to the 4th Law, we see instances where the blind or deaf, or dumb can learn to read by signs and symbols, as part of their communication process.

5th Law States;
learning conferees a set realization upon its adherents; this realization proves the fact that knowledge is Light; Learning it's engine.

6th Law states
Learning influences behaviour in a certain direction, depending on the guiding force or motives behind the educator and the accepter.

7th Law State;
Learning has a reciprocal effect upon it's own operators; the more we teach, the more we learn; and the more we learn, the more we wish to teach.

8th Law State
Learning has a multiplier effect on its operators; we grow best at what we learn, and we learn best at what we grow in.

9th Law State
Learning is abstract in nature but physical in effects, whether positively or negatively.

10th Law States;
Learning is a force powerful by expression; an influencer by impression, and weakened by depression.

11th Law State
Learning has a source but has no limits; it is not limited to age, experience, qualification, status, race; for it ends in God the creator.
With reference to the 11th Law; this is proved for in stance, when the Holy scriptures reveal God's unhappiness at the realization of man's fallibility due to his sinful inclinations.

12th Law states;
Learning must be whole so me, functional, practical, specific, realistic, sensible and purposeful; anything short of this defines harrow mindedness and an illogical approach to life.

13th Law States;
The strengths and abilities of its operators are determined by the speed; fasteners or slowness of each party; for the educator-his methods or style of impartation; while for the learner, his aptitude and cognitive skills at picking steadily.

14th Law states:-
No Learning is complete without a sense of Sacrifice, influenced by behaviour such as do cility, patience and to lerance; commitment such as time and finances; and certain ratio of understanding.

15th Law States:-
Learning is a appreciated when committed and retained to memory; recalled when necessary; achieving it's peak at the solving of a basic human need.

16th Law States
The mystery of learning starts at birth when we learn to accept life by our cries, and ends at death, when we learn to accept lifelessness by our silence.

From these laws, we may deduce learning Pmakumatically as:

$$\text{Learning} = \frac{\text{Interest x Attention}}{\text{Ability}} + \text{Time / Place}$$

$$L = \frac{\text{In x Att}}{\text{Ab}} + KTP$$

Where K is a sign of constancy of Time and place.

Pmakumatical deductions will help us;

i. To carry out analysis easily;

ii. Retain to memory by formula
iii. Understand a given law better
iv. Understand better its operational methods
v. See further chances of creating a new system from the existing one.

CONSTITUENTS OF LEARNING

The American Heritage Dictionary defines the term, constituents as "serving as part of a whole; component." My research reveals what I shall refer the constituents of inherent in learning. The constituents of learning are:

(a) Guiding principles [G. P.] (b) Practical Application [P. A.]

1. Guiding Principles {GP}: These are the theories, the hypothesis, the philosophies, the stated codes of a system's behavioural pattern. The GP expresses its benefits as stated below;

i. Helps develops correctness in learning
ii. Encourages creativeness and innovations
iii. Leads to healthy competitions
iv. Breeds excellence and achievements by less energy.

2. Practical Applications {PA}: These are the physical uses of knowledge obtained, based on the Guiding principles, in order to satisfy a basic human need. Its benefits include;

i. Solves problems practically
ii. Does not depend on abstract means, but by workable methods and techniques.
iii. Helps its user earn a livelihood
iv. Helps define realistic measures
v. Helps define needs and environmental factors
vi. Here, experience of theories are gained.

It is of worth to note that the G.P. and P. A. are complements of each, yet the P.A. has a longer lasting weight since its practical sense is expected to keep the system going; after all, principles do change. They are both measurable, reciprocal, but not without being subject to further development.

Pmakumatically.:

$$\text{Learning Constituents} = \frac{\text{G.P} \times 1{,}000\%\ \text{I}}{\text{P. A}}$$

i.e. 1,000% I is used because there's no limit to learning where I means infinitum or infinite;

yet one may exceed in abstract terms, or principle, the 100% mark; for instance, the ability to create something extraordinary from simple principles.

I shall describe diagrammatically, the philosophy behind my principles of learning constituents. This is what I shall refer to as the Learning Constituents Triangle (LCT).

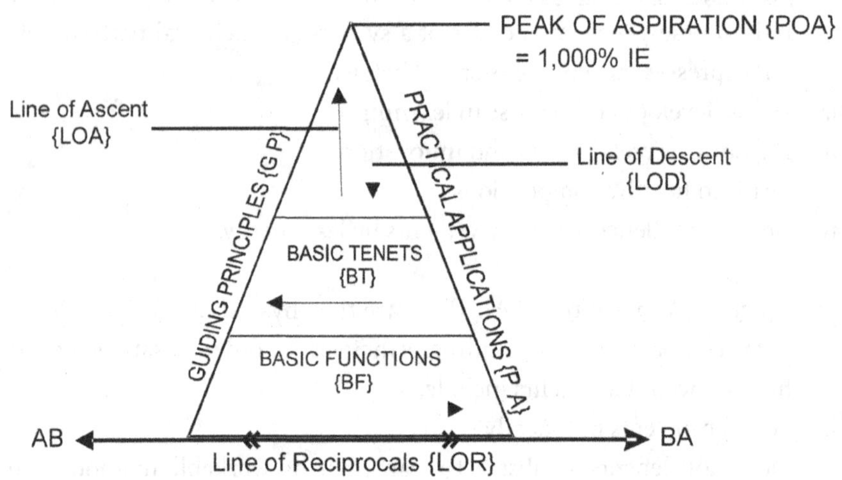

LEARNING CONSTITUENTS TRIANGLE (LCT)

Explanation of the LCT

BT is the processed rules of the GP; while the BF spell same for the PA; such that their arrows point them towards such directions. For the intellectual, they are processes represented as GP and PA; while for the lay man; processes are reduced to simple, explainable methods as BT and BF. For instance, an unlearned mechanic may be wonderful at driving yet he is a novice to simple mathematical formula such as algebra. Abstraction makes for the difference between both, i.e, when they are not complementing each other. Moreso, the BF defines the extent of one's practicality in his environment.

The lines of ascent and descent present to us the working operations of knowledge which exist at various levels. Yet it can either ascend or descend, knowledge is no flat bed. The POA is measured in 1000% I. i.e, up to infinitum (I). It is the point of the realizable wish to make a necessary impact. It is the point when the knowledge obtained yields its dividends in volume.

The lines of Reciprocals (LOR) prove that both constituents of learning complement each other as they, are equal and opposite.

2. **ADAPTABILITY**

The second purpose of Education is adaptability. Our ability to measure up in various ways, as necessary, within our earth, our world of people and most importantly, ourselves. This is the environment. Man is an independent being, uniquely created, but not isolated, i.e. he does not live alone. He is a product of influence by others. He lives in social circles or pattern, he must work with others and depend on them as well always. Education exposes him to this environment of his, but the extent of his success would depend greatly upon the level or extent of knowledge which he has. Since by this, he will be able to see, ascertain and plan in a bid to react to an encountered situation.

I have propounded certain laws, seven of them precisely to help explain better this phenomenon:

LAWS OF ADAPTABILITY

1st Law States:

The environment influences us and brings our learning to the physical test and assertion of real life.

2nd Law States:
When the environment teaches us, it does so by our experiences or that of others, which either builds or destroys our confidence, depending on our own individual mindset.

3rd Law States:
Our relevance to the environment depends on our output to its stratagems.

4th Law States:
History keeps records, whether good or bad, of our reaction and livelihood to the environment.

5th Law States:
The environment is like a mirror; what you decide to give it, it also decides to give you back, based on the principles of excellence or deterrence.

6th Law States:
Though we are products of our environment, we own the power of the mind to exceed or be limited by it; and this depends on what we hear, the company we keep, our thoughts and our actions.

7th Law States:
During learning, we use our mind the thinking faculty; in the environment, our conscience does the guiding so That we see, judge, determine and act rightly.

SUMMARY

1. The definition of education may be definitive or indefinitive.
2. Education consists of an educator and a learner
3. Factors such as consent and interest form the threads of parties in every educative process
4. Pmakumatically, education is expressed as:

$$\text{Education} = \frac{\text{Educator (ET)} + \text{Acceptor (AT)}}{\text{Time (T)} \times \text{Place (P)}}$$

OR; E = ET + AT

KTP Where K is a sign of constancy.

5. Learning is defined as acquired skill or knowledge.
6. There are processes which determine a complete learning
7. The previous standards of learning set by our fore-fathers must be complemented not destroyed.
8. Learning, creatively, consists of 16 laws;
9. The complete essence of learning is best understood by the analysis of the learning constituents triangle (LCT).
10. Education is a defeated mission without a purposeful application of its end.
11. Adaptability is guided by 7 laws.

CHAPTER 21

STARLIGHTS OF EDUCATION, FIELDS, SCHOLARSHIPS ETC

AIMS / OBJECTIVES

1. To teach the lights of illumination inherent in education
2. To expose the purity of education when it is obtained
3. To teach that creative education contains benefits known as star lights.
4. To expound on the differences between the educated illiterate and uneducated literate.
5. To discuss the challenges and solutions of education in Nigeria distinctly.
6. To discuss the benefits and methods of administering scholarships
7. To expose the student to the large world of education careers
8. To teach creative education.

STARLIGHTS OF EDUCATION

Education, whether of the Sciences, Arts or the Commercial departments or faculties ought to be pure. This would mean that it is real and practicable, and as well devoid of any bias or confusion. This is supposedly surpassing as it provides insights into the operations or working methods of a system. When this knowledge is thoroughly pursued and obtained by the acceptor, it brings a certain amount of understanding. Such understanding reveals a certain realization to the operators. This realization helps to discover by evaluation and constant re-appraisal of the facts, why things of such nature happen, its causes and effects, and further methods of improvement. This realization acts as a guide through a path; a torchlight through a dark tunnel and ultimately illuminates the formerly, once be-darkened mind of the acceptor. This is what I shall refer as the LIGHT OF ILLUMINATION (LOI).

Education is a star because it seeks to offer lights which guides its possessor towards a certain point of achievable success. It produces lights for proper vision and direction. When properly used, it becomes power. How unfortunate it is to find a nation of so many graduates who by nature are researchers, now seeking hopelessly for the jobs they refused to create. Proved and accepted your majesty; corruption and economic factors such as poverty hinder such; but agreeably, the power of governance lies with the people. No government controls everyone's mind if we dare to be creative, to help our own nation and people by our resourcefulness; rather than waste away our years of prime waiting upon the government and the public chest to draw our own sources from. Search yourself properly, the thing you search lives with you.

A person who has developed his mind properly tends to enjoy certain unlimited benefits and no one, except himself may stop it. These benefits are what I shall refer to as the

STARLIGHTS OF EDUCATION and we shall discuss these below.

STARLIGHTS OF EDUCATION

These include the following:

1. **Knowledge:**

   True education, independent of other factors grants you certain knowledge. The knowledge of yourself, the issue at hand and its relationship to the living environment is basic tool for living and earning a livelihood. The other side of knowledge is ignorance; and incomplete knowledge breeds narrowness. On the need to obtain precise knowledge, consider the following statements:

   i. The common curse of mankind is folly and ignorance William Shakespeare.
   ii. There is nothing so costly as ignorance, nothing so cheap as knowledge Ashutosh Mookerjee.
   iii. A little learning is not dangerous in itself, but may become in the modern scientific era, when the basis of life-science-presupposes exact knowledge. Mr. Arnold.
   iv. Keep your ideals high enough to inspire you and low enough to encourage you-Anonymous.

v. There is first the literature of knowledge, and secondly the literature of power. The function of the first is to teach, the function of the second is to move. De Quincy.

Knowledge is a starlight because it empowers its operators.

2. **Self-Reliance:**

This is what I shall refer to as the most powerful of all starlights because herein lies various virtues, and experience which one would ordinarily need to be responsible. A self reliant person is one who is, and acts independently, upon his own judgment and discretion. He makes his own mistakes and he also learns from them. If loyal to himself, he would strive to develop his talents and endowments, to the peak of aspiration (see learning constituents triangle). "Most happy he", said Cicero, "who is entirely self-reliant, and who centers all his requirements in himself". "The great Italian artist, Michelangelo, said "the promises of the world are for the most part vain phantoms, if we trust in these, we only delude ourselves when the time comes they invariably betray us", (D.N. Ghosh). The history of human achievement is indeed a record of what man can do by himself. Great heroes, great scientists, great merchants, they were all self-made men. Babar founded an empire relying on his strong arm.

Benjamin Franklin was born of poor parents who could give him little education, and yet, by relying on his own God-given powers he made his name memorable in science and in statesmanship. Michael Faraday began as a book binder, but rose to be one of the greatest scientists of the world. M.K.O Abiola had a poor background but lives honorably in the hearts of Nigerians. Before me, lies the history of the great Obafemi Awolowo, printed and published in one of America's greatest biography publications-current Biographies, 1987 Edition. You could do better if you believed in yourself. Call me the optimist and I will gladly answer. Education offers you freely on a platter of gold your own very source of independence-self reliance.

3. **Usefulness:**

A properly educated mind is useful to himself and to the society. He develops his mind to generate tangible results from his own initiatives.

He appreciates the importance of time, finance and other resources at his disposal. He makes proper use of his thinking faculty to solve problems. He has no time to spend on non-profitable issues such as gossips, quarrels and unnecessary celebrations. To him business comes first. He believes in good opportunities and finds them every where. He struggles to obtain rather than cringe and complain.

4. **Social Equipage:**
   Education equips one to become socially important and financially equipped. He can help himself and also be a source of inspiration to others. His education, now polished, forms an armour to fight and conquer the battle of life. That is why I disagree abruptly when it is said that education is the best legacy for a child" rather I would say "Functional Education is the best legacy for a child", such one has a say in the society, and his impact of service is thoroughly appreciated the by common populace

5. **Sense of direction:**
   Since education produces light, then it is scientifically proved that light travels faster than sound; and has direction Education is a pointer to its possessor guiding him where he should go in order for him not to miss his way. A sense of direction is a great asset because it demands thorough reasoning and the application of common sense.

By the characteristics of lights, I shall endeavour to discuss the major divisions of the light of Illumination; these I have divided into three parts and classified according to their intensity or strength of out put as;

1. Low lights of illumination (LLI):
   Characteristics of its possessor:
   i. Overtly dependent on government, friends and parents to direct him.
   ii. Personally loses confidence easily
   iii. Has no use of his common sense
   iv. Has a poor or no sense of direction

v. Cannot appropriately make intelligent decisions.
vi. Always boasting of his credentials, though makes no use of them.

2. Medium lights of illumination (MLI):
   Characteristics of its possessor:
   i. Partially blind to the application of knowledge obtained.
   ii. May sense and perceive problems, but does nothing to profer practical solutions.
   iii. Builds confidence but does not hold it reasonably.
   iv. Often good at criticisms, stuck to conventional methods but afraid to try new ideas or strategies.
   v. Loves to be recognized but unwillingly to begin little.
   vi. Lacks flexibility to expand if he tries to secure independence.

3. High lights of illumination (HLI):
   i. Is intelligent about issues, and discusses them with an open approach.
   ii. Open to new ideas and discovery.
   iii. Best when left to operate discreetly and independently.
   iv. Profers solutions with analytical and intelligent approaches.
   v. A comfortable resource he is, when entrusted with responsibilities.

CHAPTER 22

THE EDUCATED ILLITERATE OR UNEDUCATED LITERATE; WHICH DO WE PREFER?

Who is an illiterate? Who is the first instance is an illiterate? According to the American Heritage Dictionary, this is one "Having little or no formal education especially unable to read and write; unfamiliar with language and literature; ignorant of the fundamentals of a given art or branch of knowledge" Well said; well written.

But I dare state that there are noticeably, two kinds of illiteracy springing up like bubbles and striving hard to make a place through the disguise of civilization. First, the situation of certification or what I shall refer to as certificationism. Secondly, the situation where certain people, though with little basic knowledge, yet possess great talents, which when cultivated upon the fertile soils of support, would go a long way to help solve some of our basic problems such as unemployment, high cost of living due to scarcity and no competition; but are thereby denied certain basic support.

I believe it's so unfortunate that the country goes crazy for certificates by whom the owners most times may not even be able to defend when called upon to do so. During the many months of my students career analysis, it is amazing to see that over 85% of the students and adults who participated fell into the creative category with a distinctive lead over others. What happened to all those talents? What happened when a graduate of Agriculture works as a Cashier in a bank?

What happens when merit is no longer the basis for employment because examination malpractice rules the day and every wonderfully excellent result becomes a snag for suspicion, and possibly, the owner may not be able to write a 32 page sheet for an essay.

You see, we destroy our own selves. It's not educated but properly educated; it's not education but functional education. Here, everyone takes solace in blaming everyone, from the government to the people; and from the people back to the government; and the middlemen sit and enrich themselves by this game of blame. Well, what about the salaries of the teachers. I hope it has been paid; not paid but well paid. Interestingly too, perhaps, the teacher who wishes to teach might also be a carrier of the virus called "educated illiterate", especially when he must dictate or copy word from the textbook to the student. Why blame them? It is in the system.

What happened to the practical? Whose interest keeps us abreast in the education sector; is it the common good or the selfish interest? What about adopting the present global system of awarding degrees or diplomas based on creativeness and practical performance by institutions of learning? All these should be considered when making policies or regards curriculum development and general policies on education. A truly educated person is too busy with creativity and innovative research that he has no time to brag around, showing off his qualifications.

Caroline Bard writes in her book, "The Case Against College" the following statements

"A great majority of our … students who are "in college" are there because it has become the thing to do, or because college is a pleasant place to be… because it's the only way they can get parents or tax payers to support them without working at a job they don't like; because they can't get any job at all, because mother wanted them to go; or for some reason utterly irrelevant to the course of studies for which the college is supposedly organized". And then once, America also suffered this fate but as a determined said was able to return the ship back on course. As David Hapgood states on what he calls "diplomaism",

"We are well on our way to repealing the … dream of individual accomplishment and replacing it with a system in which bears no relation to performance. The career market is closing its doors to those without degrees… Diplomaism zones people into a set of categories that tends to eliminate the variety and surprise of the human experience".

He further states:

"In a system run by diplomas, all avenues to personal advancement are blocked except one: the school that gives the diploma .... When we leave the institution, like carcasses coming off a parking plant's assembly line, an anonymous hand affixes an in delible stamp ... which thereafter determines what we can do; and how we shall be rewarded. And that stamp, unlike the imprint on the side of beef, reflects neither our person value to the society, now the needs of the economic system"

According to Daniel Webster, Ph.d;

"There are at least three major problems in a society based on diplomas:

1. There may be little connection between degrees earned and on the job performance.
2. There is much evidence that vast numbers of college students are being trained for jobs that simply do not exist.
3. It way well be that the cash investment in traditional college education is an extremely poor investment indeed".

He states", whatever the reasons' the system is a confused and disarrayed one". Then carline Bird laments further", the educational requirement set by employers for jobs vary arbitrarily from region to region and fir into firm, but-educational level has nothing to do with how well the jobs are done, how well they are liked, or how long the workers stay with them".

You could consider the case of Bill gates and the micro soft story.

Then for our national good and perhaps, also personal interests, let us consider which of these forms of education at the post-secondary level is better for us; the alternative or formal education; or the Tradition/informal, as adapted from a publication of University without walls uww (Daniel Webster, PhD).

This is one of the reasons why I have written this book; to show by my personal research and wholistic views of my opinion on present circumstances surrounding our educational sector that there are other, better, more productive, more practical and functional, more sensible and more intelligent, here economical, and more efficient ways of playing the educational game, with obvious results.

CHAPTER 23

CHALLENGES OF EDUCATION IN NIGERIA

With due respect to my country and citizens, I have written on this issue based on experience and in in-depth study of the situation. However, the challenges and problems of any sector may seen, not peculiar to a specific country of a certain land mass, but in due consideration, as an educated, or perhaps an intelligent person, to the people. Interestingly, my country towers largely in economic analysis such as in the petrol sector, banking and others, far above so many counties, especially in Africa. I have after listened to analysts, whether experts or mere commentators define our problems, especially with regards to population and poverty alleviation from a myopic view of the real facts on ground. The question which first comes to mind is how the man who sits in a comfortable air conditioned car, lives in a mansion and has every necessary comfort provided for him would want to profer solutions to the seemingly looming troubles.

Learning is a painful process which requires not the wearing of suits but folding of the sleeves to go to work! What did I say? The Biblical Joseph had to be made king because he had the key to opening the door of practical solutions to the pre-determined problem; that was merit; merit earned and deserved, proved itself in worth and work; not by speaking too much grammer. When God calls one to serve, the blessings are already there as reward; but he who must serve faithfully; that's the crux of the matter.

Nigeria is a country blessed with many natural and human resources; every body knows that. Nigerian suffers many illnesses as an entity; everybody would quickly agree to that. But the Nigerian citizens are not willing to forego, sometimes, their belly and greedy tendencies, to research and make available tangible solutions to problems which continue to be-devil us.

Traveling to study oversea, should be checked against its merits and motives? Why am I wishing to travel? Is it to run away from the system to which I could make certain impacts to change for positive use and benefits of others? Consider it this way. You travel to study a specific field, such as gerontology which is, presently, not studied in Nigerian higher institutions of learning; then you return to run a sensibly affordably, functional program for the old, giving their lives a meaning and bearing fruits to yourself and society at large; or offer a certain education in fields so rare, interesting, isn't it. Nigerians are worth the salt and Nigeria is worth the benefits of this thought, if each person can research, think and work profitably, draining and employing others.

I am a true patriot and these are my thoughts. [Please see Utmost Dictionary of Careers by same Author and publisher]. No problem defies solutions if we are willing to fit ourselves into the shoes.

What have I done to profer solutions myself? See my letter on Part A. I shall, in contribution, to this analysis humbly express the solutions to these problems, in the hope that if these are not enough, then you could be inspired to add yours and do your best too. You see, you need to think about it.

What we have failed to realize is that every other sector in the nation ought to take nourishment from the education sector, or else, the results would continue to be claiming at the shaking, trembling and the never ending inflow of bad news at the streaming tide into homes and system. I predict greater down turn and sinking if no tangible considerations are made, partly upon sector as a whole. The challenges of education, in Nigerian and its simple solutions are outline below;

1. LACK OF PRACTICAL AND FUNCTIONAL PROGRAMS:

There are many programs so few are tailored to meet specific and result oriented needs. I shall particularly commend those programs such as the Cowbell Mathematics Competition, the Lagos State Government's recent introduction of the Compulsory reading hours in certain days of the week; and those programs geared towards talent discovery and youth development. But particularly, we must fight the upcoming virus in the

organization of these programs, what I shall refer to as "mineral and gifts" programs. Sincerely, these are good, but less emphasis should be placed on these, because students now want to pay, because of what feast you wish to organize, rather than what developmental benefits they would get. Education is too serious a duty and calls for every seriousness.
For our contribution, please study carefully "My Letter" on PART A: Chapter 1. I bet you will be fascinated by this program.

2. LACK OF ADEQUATE FIELDS OF STUDY AND SKILLS IMPROVEMENT:

Our research shows that Nigeria is presently not studying over a hundred courses. These careers ranging from "academic-based fields" to "vocations". But how does this affect our economic output and general progress?.

Consider, for instance, the courses, "Business of banking" and "Banking and Finance". While the first teaches one how to run a bank from the scratch; dealing technically and intelligently on issues such as long term loan management and low interest risk management, entrepreneurial and management skills, as well as responsibilities, the latter [banking & finance] restricts basically its student to the expectations of the employer from the employee, that is, how to be an employee. Presently, when the banking sector refuses to fund certain inventive programs, or the chances of such a business dying early or suffering a shock is imminent and that sector suffers, from an economy would progress or retrogress steadily upon the output of each active season.

Consider, also, a study on "Sports Administration", presently lacking in our system. Yet our country is yet to reap to the maximum, the full benefits from sports and sports enthusiasts littered all over the streets of Nigeria. No one has considerer inculcating such a course of study, professionally inclined overseas; into our tertiary curriculum. So, what results is the wave of continuous expatiation from this sector. Where we linger, we litter.

As long as education continues to suffer in this country, the nation and its many sectors shall remain its victims. Records have it, that the best medical doctors and scientists overseas have Nigerians in their nets. Why? Our medical schools suffer from lack of technical professionalism and lacks update in global trends. What about Aviation and space sciences? While the Americans, Russians and others struggle to place their landmarks; planning a raid on the possibility of livelihood on the moon, we are still doubting the discovery of more planets. Education cannot be left to the government or the sole educators alone, let credible people be called upon to deliver. Let us learn to judge based on records and results, basis for merit, rather than on the size of bank accounts and properties owned. It's our faults; it's our own faults.

For me, I have created a book titled "Utmost Dictionary of Careers" comprising local and foreign courses, their applications, occupations and vocations; as well as my sixty principles on talents and potentials, just to stimulate you and move you to do something worth the while. In the later part of this book, I shall endeavour to outline, to the best of my knowledge these courses and vocations. Diversification of the economy will be normal in order to satisfy various needs and the present undue pressure upon few careers will be reduced to the barest minimum.

3. POOR FINANCE & EDUCATIONAL ORIENTATION

Big problem it is. Every parent wishes to have their children become well educated, yet, so few really understand why their children ought to be educated. They have not realized that education is a necessity, not a luxury; a right, not a privilege; a credible investment not an end; a responsibility, not a delegation; an objective, not an alternative. Unfortunately, education has become a source of what I refer to as "Social Retaliation", a situation where a child's education is sponsored with the ulterior and superficial motives of fighting back the hostile system, using the child as a weapon. Sometimes, we hear statements such as:

I. My child should become a lawyer so as to fight for our family land;
II. My son should join the Army so as to become a terror in my family, or among my friends
III. My daughter must become an accountant since we are a family of accountants.

Social retaliation is usually the precedents or pathway leading to undue parental influence [UPI] on a child creating confusion between his actual endowments and his parents' selfish desires. He has no option but to study according to their wish, since they are his sponsors. Then he completes his study but cannot perform. What about the issue of the Child's Disobedience to Counselling (CDC). Though ill-informed, he is being guided by experts who have carefully studied his records, for this is beyond mere guess or ability, but yet he refuses to listen. Personally, I have a story to tell about myself with reference to this. My lawyer uncle Tony, once counseled me to study Arts during my senior year. Law precisely, rather than the sciences of my pride which I longed for, in order to boost my ego, please note that I was in JSS 3, then; about thirteen plus; I refused. About four years later, my aunt Mrs. Rita Akinlade an expert in Guidance and Counselling, and whom I worked with on casual basis while she was the National Youths Service Corps (NYSC) Director Delta State, advised I study Library Sciences; Guidance and Counselling on Education. I still refused because I loved adventure and would like to be a pilot. As a one-time student of Madonna University, Okija the first option offered me was a Diploma in Education Library Sciences; which I was able to reject and register how, since my lecturers and Dean appreciated my flair for then. Yet I did well in my first year of Legal studies; and would eventually break out few months later to study Aviation operations and management of which of which I still did well. Not until I became independent, that I began to see where I most fitted in. This book, my plays and other works bear testimonies to my calling to Education [see brief Biography of Author at the beginning of this book] rather than to my training. We must be loyal to our conscience first and the rest shall take care of itself. Counting on my experience, I still aspire to study both courses, but now at my own expense. This is my story, what is yours?

4. GOVERNMENT POLICIES:

Policies are measures taken to regulate and standardize the activities of its people. A good government policy should first of all consider its objective, as well as its overall infact on the people. A credible policy on education would be simple, human oriented, not profit-oriented for education yield, dividends not only to the pocket but to the environment; flexible and measurable. Issues such as tax systems, school approval requirements, practical programs, healthy competitions, incentives for outstanding performance of schools and students and the creation of the creative sciences in both the conventional school and the vocational, should be a priority; funding and support school fees regulatory measures while government supports and others; such as curriculum tailored to meet with required global and national needs such as to address global warning, food crisis, entrepreneurship, accountability, democracy studies and strict compliance to high moral and educational standards such as punctuality, practicality, discovery, observations and mental alertness, punishment of examination malpractices by a special education court and all these passed into law by the legislature.

5. LACK OF CREDIBLE PLATFORMS

It is easy to assume that something is being done to protect the right of the child to education. But I say that it is not enough. We have little or ineffective legal and finance platforms for supporting our educational sector.

What about creating special colleges for acting and stage performances, literary and skills development, creative sciences, security, sports academy (whether specific or combined); then the creation of a special ministry, apart from the Ministry of Education. Special banks to fund only this sector, and courts to try cases regarding education, or perhaps, the introduction of a course such as EDUCATION LAW, as we have COMMERCIAL LAW? These platforms should be institutionalized to ensure long term benefits.

This is a greater investment for the investor and the populace. Education, in Nigeria, still has a voice, for we hear it everyday, and its impacts are felt everywhere. This sector is an empire which confers power if we can hold it at this hour.

6. LACK OF SCHOLARSHIP FACILITIES

A careful study of global trends especially the United States, tells how this happens. Nevertheless, I am sure your mind soared when you got this book, because finance based on careful planning and application will no longer be your problem to studying overseas.

That is why this book is made. To place your one future into your own hands, and with God, you'll surely get there. True!

Interested?

Brief History about Degrees

1. "Doctor" has been a title of respect for a learned person since Biblical times [see Deuteronomy 31:28]
2. At the University of Bologna and at the university of Paris (the Sorbonne), in the mid-twelfth century, the outstanding scholars were called either "Doctor" or "Professor", or "Master", the three terms being used interchangeably.
3. The first American degree of any sort was given by Harvard University in 1642, and it was a Bachelor of Arts to the college's first graduating class, of nine men. (Harvard was a college, not a university at that time.
4. In general, "University" means a group of colleges, all under the same administration.
5. "It takes a doctor to make a doctor" has been a long tradition.
6. The first woman anywhere to earn a doctorate was probably one Novella Andrea, who lectured in law, from behind a curtain, at the University of Bologna in the 14th century. From behind a curtain? Well, she was said to be so beautiful, her students became terribly distracted when they saw her.
7. George Washington was given the first of his six or seven honorary doctorates by Harvard in 1776. (Washington thus has about a fourth as many doctorates as Bob Hope, but Herbert Hoover still leads the honorary doctorate sweepstakes with 89.)
8. The first American woman with a doctorate was M. Carey Thomas, who received one from Zurich in 1882, after being rejected by the German Universities, who did not then accept women.
9. The first black man earned a doctorate in America in 1876, Edward Bouchet, in Physics, from Yale.
10. Several black women first earned a doctorate in 1921.
11. Oberlin college in Ohio was the first American school to admit women, in the late 1830's just 200 years after Harvard admitted the first men.
12. Bachelors degrees came doing well after doctorates.

The probable origin of the term is from the old Latin "bacca" meaning "Cow", and the first "baccalaureus" was the one entrusted with the care of the cows.

The term later evolved into meaning "apprentice" at anything. But in America, it was felt by many that the term was in appropriate for a woman, so by the late 19th century, schools were giving female graduates degrees such as mistress of Polite Literature, Maid of Philosophy, and Sister of Arts.

Have you considered our latest titles, Master of Creative Sciences (M.Cs) for the man, and Mistress of Creative Science Mst; CS for the woman; Genius of Creative Sciences (Respectably) G;CS (R) and Mistress De Genius of Creative Sciences (Respectably) MDG; CS (R); and Sage of Creative Science (life for life) S; Cs (L) and Mistress De Sage of Creative Sciences (Life) MDS; CS (L)?
I bet you would be thrilled.

And so I hope that Nigerians, as well as people in general will come to regard knowledge as more important than credentials, although they still matter, but must be defended. There is no short cut to true and lasting success.

Nations have recently been led
to borrow billions for war;
no nation has over borrowed
largely for education.

_________ Flexner

The intellectual world is a market place of ideas, one without frontiers….. where the mind is not in chains and where learning does not take its bearing from the Bully king's oracle or his cannonading intimidation.

Onome Osifo-Whiskey.

CHAPTER 24

BASICS OF FINANCIAL AID

Financial aid to funding education has been an issue of topical and extensive debates over the years. While some countries have been able to work out the modalities for solving its various puzzles especially the United States and Canada, others especially the African countries have not been fair and practical enough to its ideals.

It is interesting to know that the best brains; most of who were born to poor parents or sponsors were discovered through this opportunity. The governments of the United States, Canada, Russia, China, and Japan understand the importance of providing funds for educational opportunities to those whose intentions and inclinations bend towards education. We all testify to the various achievements of these countries in science and technology, the Arts & Humanities, Medical and Pharmaceuticals, etc.

When people are creative and well-funded, the streams of great benefits spread to the general citizenry and the country.

In this section of this book, we would try to discuss and discover what financial aid is and how it is operated, and perhaps see the positive reality of obtaining money from the right sources. It is also necessary to have an open mind in order to solve the right problems especially in such issues which bother on financial aid.

CHAPTER 25

WHAT IS FINANCIAL AID

In simple terms, the word "Financial" refers to monetary resources; funds, its management, banking and investments and credit. It also means to provide money or funds for a need. The word "Aid", means to help; support, assistance.

From those, we may define Financial Aid as a money resource raised to help, support or offer assistance to those who need it urgently. In this situation, the sources recognize through certain means of enquiring the need to offer such help financially to the student or applicant. However, an applicant who requires financial aid must be able to prove this. Such an application, more often requires certain information such as personal and parents (family) background, educational background previously and presently, marital status such as single, married, widowed, divorced, separated, engaged; age and date of birth, nationality, hobbies, motivation and personal tastes and attitudes.

Interestingly, financial aid is wrongly assumed for the poor, but most times, it is the supposedly rich who eventually receive more aid. Yet, it is proved that there is money for the purpose of educational development the world over.

CHAPTER 26

SOURCES OF FINANCIAL AID

The sources of financial aid vary. It could come from local sources such as sponsors, philanthropists or from foreign sources depending on one's academic commitment. Nevertheless, it must be understood that there are many organizations which have been set up for this purpose of offering financial aid to deserving ones. A major condition for obtaining finance to study is that an application is made after understanding the basic requirements of such an organization. These information may be obtained from the office of admissions (if University); office of secretary (if company); pamphlets and bills (at seminars); and on the internet.

The sources of financial aid include private businesses, research institutes, international and multinational companies, foundations, universities, humanitarian organizations, private organizations, religious organizations, government.

Based on numerous studies, a great number of scholarships are not used simply because parents or students do not know where to apply for financial aid. Many studies have concluded that millions of dollars in scholarships from these sources are never used. Based on research, many students decide not to pursue a university degree because they don't know that there is money available.

CHAPTER 27

CATEGORIES OF FINANCIAL AID

The categories of financial aid are dependent on the situation appropriate to the student or applicant. It is divided into two kinds of classifications.

(a) Types (b) Classes

A. The types of financial aid are:

(i) Need based aid (ii) Merit based aid

NEED BASED AID: This is the financial aid offered to a student applicant because of under-funding. Most aid is developed on the basis of eliminating or at least, reducing what term "educational poverty" to the barest minimum.

MERIT BASED AID: This is offered to students of outstanding intelligence, brilliance, and excellent performance. They are offered this aid because they have shown intellectual promise through their performance in a certain field. They are believed to deserve this and their family income is not a determinant factor.

B. Classes of Financial Aid:

The classes of financial aid to education are broadly divided into (a) Scholarships (b) Loans (c) Grants (d) Fellowships (e) Exchanges (f) Lotteries and Mojo

Their features are summarized below:

(a) <u>Scholarships</u>:

Definition: A grant awarded to a student.

Features:

1. Money is awarded based on a defined type
2. The student does not repay after studies

3. The student must fulfill certain requirements of the awarding institution or organization
4. It is not limited in scope
5. Applications and awards may come from various sources
6. The student's study interest level and motivation must be high
7. Scholarships do take time since application date and review date quite differ.
8. Exists in local and international awards.

(b) Loans:

Definition: A sum of money lent to the student applicant for a period of time, to be paid back to the loaner with interest after studies.

Features:

1. Money lent by the student is expected to be paid back with interest
2. Has a definite time frame
3. Sometimes; it requires a co-signor, a guarantee, a collateral or security
4. Exists in variety in order to cover student's needs e.g. insurance, accommodation, tuition etc
5. Subject to economic factors and negotiations between both parties
6. Subject to legal scrutiny, for instance, in times of bankruptcy, it is expected that the debtor files for bankruptcy at the court of competent jurisdiction.

(c) Grants:

Definition: A giving of funds to a student applicant for the purpose of studying. It is the same in principle with scholarships except that they differ in name or title.

Features: Its features are the same with the scholarships.

(d) Fellowships:

Definition: (a) The financial grant made to a fellow in a College or University

(b) The status of having being awarded such a grant

Features:

1. Most often, it is discriminatory in nature as it is offered to only students of a particular institution by the same institution.
2. It involves scrutiny and close observation of such students.
3. Certain points or grade points must be attained in order to receive this award.

(e) **Exchanges**:

Definition: Also referred to as Exchange programs. It involves an institutional transfer of students, between party institution, often to study in different regions.

Features:

1. Designed, most often, to enhance research by students
2. Involves an agreement and sharing of interests by party institutions.
3. Often cuts across national and international boundaries.
4. Budgets to cover all expenses for the participating students are made
5. Involves, like other awards, a certain umber of people
6. Highly restrictive to certain fields of study which most often are perceived new and unknown terrains, for instance, an International Exchange program involving American students to study certain features of African cultural, agricultural or business terrain and development prospects; and as well the same number of African students doing the same over there.

(f) **Lotteries & Mojos**:

Definition: A contest in which winners are selected in a drawing of lots. Often referred to as Mojos by Colleges, especially in the United States and Canada.

Features:

1. Often a contest with basic rules of observance.
2. Often involves a certain test of ability, and knowledge of specified fields by its organizers.
3. Limited to few winners who win bulky amounts.
4. Involves an open invitation, like others, to participate.
5. Has expiring dates.
6. Lots are drawn by criteria and standard of works.

SCOPE OF FINANCIAL AID

Briefly, while some are limited to sponsoring to a certain extent, others are non-limited. it is necessary however to study the conditions of the awarding institutions.

CHAPTER 28

BENEFITS OF FINANCIAL AID

Financial aid, when well administered has certain benefits.
These benefits are:

1. Discovery: It sure does lead to the discovery of the best brains, both academically and vocationally.
2. Development: Companies and private organizations, who have sent their staff on scholarships would readily testify of the extent of developments which these ones have contributed on return
3. Opportunities: Where there are scholarships, there are opportunities. Opportunities for the less fortunate, less-privileged and less provided for. Most times, such ones become great beyond our very imagination.
4. Training: Scholarships provide the much needed training required to develop and utilize specialization in every field and skill.
5. Efficiency: Financial aid, definitely promotes a healthy spirit of competition and thereby leading to efficiency.
6. Value: Financial aid respects the hidden endowments on its beneficiaries and thereby places value on the persons.
7. Appreciation: Most beneficiaries of financial aid have come to the appreciation of life, develop a sensible sense of personal esteem, bred confidence and gave their best to their calling and careers.
8. National Development: Most great nations of the world realizably, could not have attained such heights without a comprehensive government support for financial aid.
9. New Studies: In Nigeria today, certain areas have not been properly harnessed. With a comprehensive financial aid plan, citizens could venture, explore and study new fields, creating opportunities and employments where necessary.
10. Research: There is no end to education. Despite leaving school, so financial aid could enhance private developmental research for human use.

CHAPTER 29

METHODS OF ADMINISTERING SCHOLARSHIPS

1. Responsibility: Government must consider it part of their responsibility to encourage and develop methods of offering financial aid to eligible and interested persons. While organizations should see it as part of their corporate social responsibility. The government support to such companies could be in the form of tax cuts, financial support and other methods as may be considered fit.

2. Goodwill: Financial aid must be considered in the light of goodwill; from personal, private and government support to the open society of the needy but yet resourceful persons. While the society must give due recognitions to those ones as service to humanity.

CHAPTER 30

FINANCIAL AID INFORMATION

Financial aid information is necessary in order to ensure that an applicant knows its many benefits as well as how to apply and obtain such by the standard requirements.

The sources of financial aid information may be obtained from newspapers, Magazines, radio and television broadcasts, embassies, corporate and consultancy organizations, internets, migration experts, professional bodies, specialized books and through certain bulletins as may be concerned with educational development as well as books and resource persons.

Information is the key to knowledge and the application of this knowledge is bound to generate waves and developmental strategies. It is therefore considered, highly necessary for students to be aware of the enormous opportunity created for their development through international financial aid sources. These sources include the United Nations Organization, United States (Private and public sectors), Canada, Universities, Corporate institutions and philanthropists. It is interesting to know that financial aid has existed in the United States for instance, for over fifty years. Others are Russia, Japan, China, etc. Where there is an interest, there are existing strategies and these are left to the interested individual.

CHAPTER 31

COST EVALUATION

There is noting good which does not place a certain monetary value on it. Erroneously, people in terms of parents and students who have expressed interest in requesting and applying for financial aid have often assumed that there is no need for money. It is however necessary to advise that education, even though it should be made affordable ought to be viewed in terms of its far-weighing benefits and empowerments, rather than the amount of money associated with it. Proper cost evaluation should be a relative comparison of the specific benefits to be obtained from such an education to the amount it would cost; then the quality of such education. The most expensive is not always the best but the quality determines the best.

Before applying for a scholarship, especially as an international student, it often costs more the applying as a local person or citizen. International application would require a cost analysis of the required examination to be taken (e.g. TOETL, SAT, GMAT); passports, correspondence, mailing and documentation. The United States and Canadian scholarships often cater for flight expenses, tuition, accommodation, migration costs and living costs. Yet, there is need to apply to as many scholarships as possible since only one will not be able to cover all your needs. Remember, it is a contributory process out there.

CHAPTER 32

ADMISSIONS

Obtaining an admission will most often depend on your choice of school. However, it is important to follow the guidelines stated below:

1. **Choose a School:**
   It is necessary to do a Virtual Tour or visiting of many schools on-line as possible. This will help you make up your mind on which to attend. It is also important that you have your factors well determined ahead of time. These factors, dully considered will include:
   - The criteria for admission [i.e. the general and specific requirements)
   - The field and level of your intended studies [e.g. undergraduate, graduate, post graduate etc)
   - The school's location
   - The site and size of the establishment
   - The facilities and activities offered by the establishment.

2. **Write a Letter of Enquiry:**
   Write a letter to your choice schools, making enquiries concerning issues such as your field of study, level, international student status, accreditation of your papers, admission exams, immigration, accommodation and housing and their scholarship facilities. Do not forget to indicate your name, contact address, phone numbers and e-mail address, at the end of your letter.

3. **Admission Exams**
   Research shows that most of the schools in the USA require you take a language proficiency exam called the TEST OF ENGLISH AS A FOREIGN LANGUAGE [TOEFL] If so, then it is important to contact experts on this in your area of residence. The local representatives

must give to you the cost of registration, practical classes, text kits [Books and materials] and must record your scores. These scores must be forwarded to your intended school of study for admission. But it is important to enquire first of the school's requirement.

4. **Conduct on-line search**

For financial aid, it is necessary to conduct a basic research concerning your eligibility to receive scholarships. Study the various requirements of each eligible organization and write them a "letter of Interest" in order to send you forms, which you must fill and return by email, fax or mail. Most organizations have deadlines after which your application would be considered late. Be sure to note this and apply early.

Applications for scholarships are reviewed from the date of deadline. This often leads to some delay as various organizations do have different deadlines. While some organizations might acknowledge reception of your letter or application others might not. Do not let this worry. The format of payment shall be communicated to you by the awarding organizations. Some scholarships are renewable annually while some are not.

CHAPTER 33

FIELDS COVERED IN SCHOLARSHIPS

The word scholarship as defined by the American Heritage Dictionary is "A grant awarded to a student", the Collins Gem English Dictionary says, "Financial aid given to a student because of academic merit". The good news here is that the aid covers both vocational and academic based fields. They are easier if you have been assessed on SCAP, for recommendations and records. Please see Part A, Chapter 1.

These fields are listed below:

A. Engineering:

1. **Aeronautical Engineering**
2. Aerospace Engineering
3. Systems analysis
4. Cybernetics
5. Pre-engineering
6. Manufacturing
7. Agricultural Engineering
8. Nuclear Engineering
9. Engineering technologies
10. Engineering Science
11. Computer Science
12. Information Science
13. Programming
14. Systems Engineering
15. Petroleum Engineering
16. Mining Engineering
17. Materials Engineering
18. Electronics Engineering
19. Management Information Systems

20. Physics Engineering
21. Bio Engineering
22. Biomedical Engineering
23. Chemical Engineering
24. Civil Engineering
25. Electrical Engineering
26. Environment Health Engineering
27. Industrial Engineering
28. Computer Engineering
29. Mechanical Engineering
30. Ocean Engineering

B. Medicals/ Sciences/ Pharmaceuticals

1. **Audiology**
2. Art therapy
3. Cardiology
4. Dietetics
5. Sonography
6. Occupational therapy
7. Premedicine
8. Predentistry
9. Preveterinary
10. Preoptometry
11. Prepharmacy
12. Biomedical Sciences
13. Food & Dairy Sciences
14. Clinical Laboratory Science
15. Medical Laboratory Science
16. Nursing
17. Medical Laboratory Technologies
18. Recreational therapeutic
19. Food processing
20. Learning disability
21. Communication disorder
22. Speech pathology
23. Pharmacology

24. Physical therapy
25. Radiology
26. Occupational health
27. Health and Fitness
28. Exercise Sciences
29. Orthotics & Prosthetics
30. Dental Hygiene
31. Respiratory therapy
32. Sign language
33. Dentistry
34. Medicine
35. Nuclear Medicine
36. Sports Medicine
37. Emergency Medicine
38. Veterinary Medicine
39. Music therapy
40. Neurosciences
41. Food science
42. Optometry

C. Language / Art Studies

1. English studies
2. Translation
3. Linguistics
4. Comparative Literature
5. Creative Writing
6. Speech & Rhetoric
7. Foreign languages

D. Agric./ Earth/ Sciences

1. Agriculture
2. Agronomy
3. Astronomy
4. Astrophysics
5. Forest Production
6. Conservation

7. Plant Protection
8. Fisheries
9. Life Sciences
10. Environmental Science
11. Marine Science
12. Soil Sciences
13. Earth Sciences
14. Poultry
15. Botany
16. Marine Biology
17. Ecology
18. Ecotoxicology
19. Entomology
20. Forestry
21. Geochemistry
22. Geography
23. Geology
24. Geophysics
25. Horticulture
26. Meteorology
27. Oceanography
28. Zoology
29. Biochemistry
30. Biology
31. Applied Physics
32. Materials Science
33. Science Technologies
34. Marine Biology
35. Chemistry
36. Mathematics
37. Microbiology
38. Physics

E. Arts/ Humanities

1. Archaeology
2. Anthropology

3. Policy Analysis
4. Law Enforcement Administration
5. Law
6. Criminology
7. Communication
8. Biopsychology
9. Library Sciences
10. Liberal arts & Humanities
11. Liberal arts
12. Education
13. Ethics & Society
14. Area studies
15. Pre-law
16. Gerontology
17. History
18. Journalism
19. Communication media: Radio & Television
20. Philosophy
21. Psychology
22. Telecommunication
23. Sociology
24. Public Services
25. Human Services
26. Political Science
27. Military Science
28. Religion
29. Public Relations
30. Advertising
31. Social Work
32. Theology

F. Finance/ Administration/ Management

1. Business of Banking
2. International Business
3. Economics
4. Accounting

5. Sports Administration
6. Business Administration
7. Actuarial Science
8. Home Economics
9. Entrepreneurship
10. Media Studies
11. Finance
12. Agribusiness
13. Arts Management
14. Wildlife Management
15. Business Management
16. Aviation Management
17. Industrial Management
18. Management Information Systems
19. Facilities Management
20. Turf Management
21. Hotel Trade
22. Real Estate Business
23. Marketing
24. Actuarial Mathematics
25. Quantitative Methods
26. International Relations
27. Human Resources
28. Management Science
29. Secretarial Services
30. Tourism
31. Banking and Finance

G. Vocations

1. Landscaping
2. Architecture
3. Graphic Design
4. Studio Arts
5. Fiber Arts
6. Jewelry
7. Ceramics

8. Film and Cinema
9. Dance
10. Music
11. Musicology
12. Computer Graphics
13. History and Conservation
14. Graphics & Printing
15. Design
16. Screen Writing
17. Fashion Design
18. Drawing
19. Interior Design
20. Painting
21. Photography
22. Sculpture
23. Printmaking
24. Theatre
25. Music Theory and Composition
26. Urban Planning
27. Voice and Choral
28. Art History & Conservation

- For courses and vocations meanings and applications, please order the book: Utmost Dictionary of Careers by same Author.

THE JOY OF AID

If there is a bright side to all this, it is that despite all the cuts, there is still a great deal of financial aid available and we are talking billions of dollars. At these.... almost every family now qualifies for some form of assistance.

Who gets the most Financial Aid?
You might think that the families who receive the most financial aid would be the families with the most need. In fact, this is not necessarily true. The people who receive the most aid are the people who best understand the aid process.

Is this Legal?

You bet. All of the strategies we have discussed in this book follow the law to the letter.

Is this Ethical?

Parents who understand these rules get the maximum amount of financial aid they are entitled to under the law. No more, and no less.

Is this only for rich people?

Many people think that ... financial strategies are only for millionaires. In some ways, they have a point: certainly it is the rich who can reap the greatest benefits. But financial aid strategy is for everyone. Whether you are just getting by or are reasonably well off, you still want to maximize your aid eligibility.

The Scholarships Bottom Line

Of course, there is a great deal to this service; there is the value of tradition, the exploration of new ideas, the opportunity to think about important issues, the chance to develop friendships that will last for a student's life time.

Some of the aid strategies are complicated, and because we do not know the specifics of your financial situations it is impossible for us to give anything but general advice. Nor can we cover every eventuality. We recommend that you consult with a competent professional about your specific situation before proceeding with a particular strategy.

The process usually lasts between 8 12 months. But your determination can work for you if you follow through carefully. Do as much as possible to keep your general information to your belt and tell no lies. It is not about lying to reap off the system, it is about answering questions simply as required. Do not reveal that you have applied to any other source to get the maximum aid. Remember, one scholarship is not enough to fund all expenses, so you must apply to as many as possible.

I wish you the best on this discovery. Good Luck!

SUMMARY

1. Education is not limited to mere studies but applicability also
2. Pure education confers a certain sense of realization called the light of illumination.
3. The benefits of education is called the star lights. These starlights include knowledge, self reliance, usefulness, social equipage and a sense of belonging.
4. The light of illumination has levels and characteristics applicable to every person.
5. Every society must define its education in terms of the educated illiterate or uneducated literate.
6. Creative education recognizes all the opportunities available in traditional and alternative education.
7. Challenges of education, is not specific to Nigeria alone, but every where.
8. Challenges of education include:
   - Lack of practical and functional programs
   - Lack of adequate fields of study and skills improvement, poor finances and educational orientation
   - Unfriendly government polices
   - Lack of credible platforms, lack of scholarship facilities.
9. Financial aid is about providing opportunities to study for those who cannot afford it.
10. Financial aid includes scholarships, grants, loans, lotteries and mojos, exchanges and fellowships.
11. Proper research is necessary to find sources of financial aid.
12. The opportunities provided by the financial aid process is a trigger to economic development
13. The development of more courses of study should be based on useful skills
14. When the educational system of a nation is not expanded enough there is tremendous pressure created on the existing few, leading to inefficiency and pressure in the labour market
15. Education is the basic, provably, foundation of every sector in the nation

16. When the products of educational scholarships are half-baked, then, invariably, there would be half-baked workers, unprofessional, non-creative and lacking in marketable skills.
17. The greatest stimulant for continuous development in any society is research
18. Research is necessity, not an option in every sector of an economy.
19. Productive research enhances and updates the status of its people.
20. Purposeful research makes the great difference between the educated illiterate, uneducated literate and the qualified literate.

Education is a field of play;
the quality of your play
necessitates the strength of your score.

Education is a shelter from the rain;
Education is a shade from the sun's blaze;
A guide to reach the point
A shield against every storm.

"The bread soaked from sweats
here, feeds our men;
the threads of the seamstress
there clothe our women;
many designs upon our minds
still please our children;
Oh! if only we could teach them
secrets of the creative minds.

SECTION 3:

Creative Entrepreneurship (The Business Aspect)

CHAPTER 34

WHAT IS ENTREPRENEURSHIP?

Aims / Objectives:

1. To help the student learn, acquire and apply his creative skills to do profitable business.
2. To reach the relationship between entrepreneurship and creativity.
3. To expound on the true meaning of creative entrepreneurship.
4. To discuss the various factors and laws which lead to self reliance and confidence.

What is Entrepreneurship?

The concept of entrepreneurship is one which has fascinated the scholars of management over the years, taking into consideration the various schools of thoughts, inherent in the ever widening field of management and its systems of operations. Nevertheless, for the purpose of this study, we would want to discuss certain basic concepts of entrepreneurship in order to equip through very simple outlines, what is obtainable in this process.

The American Heritage Dictionary, third Edition defines entrepreneurship as "A person who organizes, operates, and assumes the risk for a business venture". This in very simple terms, projects an entrepreneurship as a person responsible for his own business. In other words, he is responsible for producing and making available for sale, on demand, his various goods to the people or his market.

Relationship between Entrepreneurship and Creativity:

Entrepreneurship and creativity exist together in a special force of reciprocals. This implies that a creative person has his resources from his internal or inside bowels of rich abundant potentials and must understand the simple methods of selling this to the world in order to obtain an income. I shall express this principle as the entrepreneur's law of reciprocals.

Creative Entrepreneur's 1st Law of reciprocals states that "the forces of creativity and entrepreneurship have equal and opposite effects upon each other".

Pmakumatically expressed:-

C.E.L. of R = Creativity Entrepreneurship

This in statement means that the entrepreneur, who knows how to improve and manage his internal resources, in order to offer certain valuable services in great demand, would definitely succeed upon his lots.

Creative Entrepreneur's 2nd Law State that: "The creative entrepreneur can only give to the market to the extent of his developmental capacity". This is the Law of availability. It implies that his resources dry up or remain active as long as his will asserts, accepts or rejects this.

Mathematically expressed as:-

C. E. L. of AV = C.E. market.

This law, however, places a great responsibility upon such a person to find ways of innovations and continue to be in place. His ideas must be defined, tested and specialized to specific needs if he must.

Pmakumatically expressed as:-

C.E.L. of Ur = C.E. SOCIETY
Where A.O.T. represents "Ahead of Time"

CATEGORIES OF CREATIVE ENTREPRENEURS

It is important to note, here, that every creator is an entrepreneur, but not every entrepreneur is a creator. An instance, here, would be a business man whose venture lies in importation or exportation, or general marketing of what others produce. He is not the creator, or inventor of such product so he is not the patent right owner. Yet, the creator creates more work for this purpose of selling.

In this light, we shall distinguish the three kinds of creative entrepreneurs.

1. **Scientific:** These include the technologists, inventors, medical druggists (one whom I shall describe as specialists in drug making), architects, etc.
2. **Literary**: These include writers, artists, project makers e.g. concerts of special purposes, musicians, cartoonists, drawers, sculptures etc
3. **Li-Tech**: These comprise people who are multi-gifted. They many work in any category e.g. a person who writes a science fiction; or is able to organize a legal understanding by a mathematical rule become relevant in the scheme of things.

Creative Entrepreneur's 3rd Law: states that "The creative entrepreneur has the special ability to predict what needs are most demanded by simply processing certain information and foresight and the effects if not present". This is called the law of predictability. He more or less can determine what society would become in a few days, or months, or years, if something is done about something or if nothing is done about something.

Pmakumatically expressed as:

C.E.L. of P = C.E. SOCIETY (1). (0)

Where (1) represents something of something,
And (0) represents nothing of something.

By this principle, he personally sensors the standard of living in the society and acts quickly in support or against a particular trend.

Creative Entrepreneur's 4th Law states: "The creative entrepreneur does not work by time rather he works ahead of time, and must act quickly as though his resolution is weakened by delay". This is called the law of URGENCY. He must bring to pass what is not, yet, in existence or modify it to become useful.

SUMMARY

1. Entrepreneurship is the process of organizing, operating and assuming the risk for a business venture.
2. A person who engages or practices entrepreneurship is called an entrepreneur.
3. Entrepreneurship and creativity exist together in a relationship of laws: the 1st law of Reciprocales, the 2nd law of Availability; the 3rd law of Predictability; and the 4th law of Urgency.
4. The 1st law or reciprocates states that the forces of creativity and entrepreneurship have equal and opposite effect upon each other.
5. The 2nd law of availability states that the creative entrepreneur can only give to the market to the extent of his developmental capacity.
6. The 3rd law of predictability states that the creative entrepreneur has the special ability to predict what needs are most demanded by simply processing certain information and foresight and the effects if not present.
7. The 4th law of urgency states that the creative entrepreneur does not work by time, rather he works ahead of time, and must act quickly as though his resolutions is weakened by delay.
8. Every creator is an entrepreneur but not every entrepreneur is a creator.
9. Creative entrepreneurs exist in the three categories of (a) Scientific e.g. technologists, inventors etc; (b) Literary e.g. artists, musicians, sculptors etc (c) Li-tech e.g. a literary technologist, a writer and printer; a creative account; as combined.
10. Each law of creative entrepreneurship is expressed Pmakumatically.

CHAPTER 35

THE BUSINESS ASPECT OF CREATIVITY.

Aims / Objectives:

1. To teach the students the basic principles of creative networking achieving outstanding financial independence.
2. To impart into the student the cardinal points to be looked out for in creative success.

THE BUSINESS ASPECT OF CREATIVITY.

It is a bad creator who creates, but cannot sell. Your creative success will depend chiefly on its ability to sell you by selling itself. The essence of creative endowments is to profit you reasonably. The creator centres all requirements in himself while yet remaining the centre of attraction Selling your works means finding a suitable market to sell your skills. A happy customer is the best advertisement you can have. You can satisfy your inner yearnings by considering my own side of your business life with regards to creativity. Let us consider what I shall refer in practical terms the following:

The Multiplication Principle:

I shall also refer to this principle as the multiple law. P. M. Aku's multiple law (ML) states that:

"Creative success does not depend on the number of days worked but chiefly on the number of copies sold and controlled by the invested finance.

Pmakumatically, NO OF COPIES MADE & SOLD X INVESTED FINANCE, but smart work demands networking which is a series of business links to yield finance so; the multiple law would be re-defined as:

P. M. Aku's M.L = COPIES MADE & SOLD (CMS) is inversely proportional ( ) to the No of Days worked, multiplied by the invested finance upon the bedrock or foundation of network; so; $\frac{\text{C.M.S. D.W}}{\text{NT.W}}$ x I. F. = C. S.

This implies understanding the issue of <u>bulk production</u>, for in printing matters, the more you produce and reproduce in greater quantity, the less you spend or invest and earn more.

Your <u>market</u> is that segment or sector of the human populace for whom you have created your work. This is in respect to certain strategic advantages e.g. geographical location, population, sex and gender, age etc. Interestingly, here in Nigeria, our high population density is a comparative advantage for a creator. By these you can be launched into creative success.

CARDINAL POINTS OF CREATIVE SUCCESS

I shall define creative success (C.S) in cardinal terms. What I shall refer to as the P.M. Aku's PODIUM OF CREATIVE SUCCESS much later. Creative success implies creative growth according to the principles of networking. This growth is four sided:-

1. <u>North</u>: Growth in Finance or Financial Growth (F.G) which is usually upwards
2. <u>South</u> (s): Downwards, Foundational skills Growth (F.S.G) is noticed. This is the process where the creator seeks to develop something greater than his previous works. He desires more from the treston the mind's own foundation.
3. <u>West (W)</u>: West wards; fame and recognition travel, locating him as though he were missing from the earth's terrain all the while; and then he becomes recognized and enjoys the fruits of fame. The F/R is the creator's shining star, the life and point of standing ovation and reception of awards.
4. <u>East (E)</u>: East wards; he becomes a positive influence (P+I) on others. Now, a shining example, he develops in sharing his time and resources by teaching skills development, poverty reduction, expansion, and capacity enhancement, trust in built and contracts are awarded him

on merit of his achievement and works. He is now a senior citizen of the world.

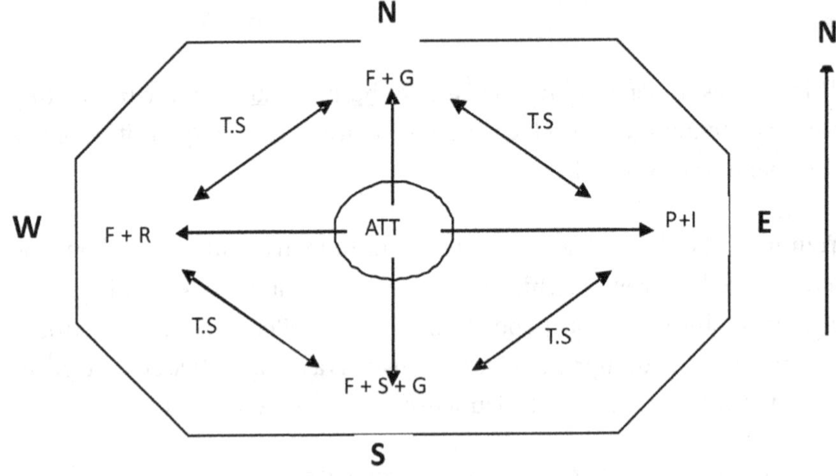

P. M. AKU'S PODIUM OF CREATIVE SUCCESS (P. M. AKU'S P.C.S.)

The cardinal points of creative success are not possible without the proper application of <u>TIME</u> and <u>STRATEGY</u> (T.S). These two deciding factors have no limits to their applicability, influencing the growth of all capacities. They are connecting, reversible and all encompassing. Time is a flowing river, while strategy is the enabling power, you drink fresh waters from time and then obtain strategic power by thinking right. The centre of the Podium is the Central Power House (CPH), of the creator's mind. It is the <u>ATTITUDE</u> (ATT). It is the central determinant, the focus, the source and depends on the conviction of the creator's mind.

The attitude is the thinking pattern (positive or negative) of the creator. Is he success oriented or environmental driven? His attitude determines not only his altitude but his limits. When success says "yes" the environment says "no" He who follows the "yes" from within his conscience shall obtain many "yeses" from without in a short time.

Though successful men have suggested the phrase "work smart not hard;" I shall rather say that creative success is the result of both working hard and working smart. Why? Energy is expended in every labour even if it's a minor telephone call and should be duly compensated and paid for. So, work hard at creating, then work smart at selling and networking because they are unbreakable bonds.

SUMMARY

1. You sell yourself by selling a good product.
2. P. M. Aku's multiple law states that creative success does not depend on the number of days worked but chiefly on the number of copies sold and controlled by the invested finance;

Pmakumatically expressed as: C.M.S. D.W x I. F. = C. S.
NT.W

Where CMS = Copies made and sold;
= Sign of inverse proportionality
D.W = Number of Days worked;
I.F = Invested finance
NT.W = Extent of Net working

3. The principle of bulk production is a determinant factor in less cost of production; the more copies or pieces made, the cheaper the production cost.
4. Market is that segment, or audience or sector of the human population for whom you have made your work.
5. The cardinal points of success is best understood by the study of the P.M. Aku's Podium of creative success.
6. The podium of creative success comprises four cardinal points:
   - North: An upward growth in finance (F + G)
   - South: Downwards; foundational skills growth (F + S + G)
   - East: Eastwards; positive influence (P + I)
   - West: Westwards; fame and recognition (F + R)
7. The cardinal points of creative success is controlled by Time and Strategy (T.S); and determined by the internal attitude of the creative entrepreneur (ATT).

CHAPTERS 36-42

Opportunity, Freedom, Democracy, Decision, Expansion, Power and Planning

Aims/ Objectives

1. To discuss and impart into the student the necessary simple yet building factors for every creative entrepreneur.
2. To teach the student that these factors individually have a link to the other, and collectively are a force for building creators for life's business challenges.

CHAPTER 36

OPPORTUNITY

The word opportunity is defined by the American Heritage Dictionary as "a favourable or advantageous circumstance or combination of circumstances", while an opportunist is one who takes advantage of any opportunity to achieve an end, often with no regard for principles or consequences.

A creative entrepreneur must search for opportunity to express his abilities all the time. An impression without expression sure leads to depression. His search for opportunity must be based on the principle made by Carlyle, "Do that which lies nearest thee, which thou knowest to be a duty. The second duty will always become clearer" The first port of opportunity is in recognizing one's own endowment. Then seeing this, he would begin to develop it along the lines of knowledge and experience. Society's needs would be clearer and then the demand and eventual satisfaction of this demand would come.

Below are steps to recognizing opportunities.

1. **INFORMATION**:
   Daily, streams of information from books, news and conversation flow into our ears. While some are positive, others may be negative. The creative entrepreneur must keep alert on the needs and problems present at the time, being in force. Information becomes useful when it is processed by analysis to develop into an idea, an impression for change. An instance here would be positive information which pertain to breaking of records in sports or business; or a negative news on earthquakes or war. Here, the creative entrepreneur would see an opportunity to cartoon his mental impression; if he were a cartoonist; or he would develop a play, novel, essay with such as his underlying theme; or he would develop theories, formulae for synthesis

upon peaceful purposes which would be sold and would stand as his work. His opportunity here would be complete upon research and development as well as his message.

2. **LEGAL STATUS:**

Legality refers to one's standing in the perception of the law. It is important that the creative entrepreneur registers a business name with the appropriate authorities in order to be allowed formally and as well not regarded as fraudulent. Secondly, he must patent his idea or what I shall refer to as copyright induction in order to enjoy a certain measure of protection and proceeds for his work and its sales. Thirdly, he must know the rights and limitations of his expressiveness. What he does is against no one but for humanity.

CHAPTER 37

FREEDOM

Freedom is defined as the condition of being free; political independence; possession of civil rights; ease of movement; frankness or boldness; unrestricted use or access. This led to the term "free enterprise", which means the freedom of private businesses to operate competitively for profit with minimal government regulation.

The first essence of creative entrepreneurship is to gain freedom, to act privately, discreetly and independently. He wishes to account for his profits and losses. He wishes to express his intentions through his works. An aspiring creative entrepreneur must define his motivation for freedom. Is it finance motivated; or a pressing need for fulfillment that inner sense of self-satisfaction which comes from the realization that he is contributing positively to his economy.

However, he must recognize the enormous responsibilities which come from being independent. Freedom is an egg shell of risks and responsibilities. Nevertheless, his freedom to act upon his individual internal resources also presents him the chances of achieving great success in life. The creative entrepreneur's law of freedom states that: Responsibility and development is the first law of creative freedom.

CHAPTER 38

DEMOCRACY

The true practice of democracy as a form of government brings many good tidings to the creative entrepreneur. Democracy is a government of the people, for the people and by the people. The people are the citizens of that country. They elect the government into power. It therefore places their opinions, views and expressions in high esteem. These people are creatively endowed and so become indispensable, by the ethics of creativity, in such government, in so much that they must lead the pace in national development.

Consider the following statement on the great scientist Albert Einstein, made in the book, AMERICAIS. "When Albert Einstein was 26 years old, he wrote several papers for a German science magazine. In them, Einstein explained his theories about space and time, light, and energy. His equation, $E = MC^2$ (energy equals matter times the speed of light squared) is the key stone to the modern concept of the atom. In 1915, Einstein wrote another paper. This one was about his general theory of relativity. This theory changed ideas about gravity held since Sir Isaac Newton's time in the late 1600's. Einstein received the 1921 Nobel Prize in Physics for his theories on light

During the 1930's, scientists began to apply Einstein's theories. They discovered how to break the atom into parts, giving off energy. In 1939, while working at the institute for Advanced Study at Princeton, New Jersey, Einstein wrote a letter to President Franklin Roosevelt. He stressed the importance of research on a nuclear bomb. He warned Roosevelt that it was important to build such a bomb and that the Nazi government in Germany was probably already working to make one.

During World War II Germany failed to make a nuclear bomb, but the United States was successful Einstein regretted his work on the bomb after he heard of the loss of life and the destruction caused by the bombs in Japan in 1945. Until his death in 1955, Einstein spoke to people and wrote articles about the continuing dangers of nuclear war".

The statement implies the right of expression which encouraged the spirit of patriotism. It is important that persons who find themselves in governance use the "opportunity of democratic accommodation" to appraise every other resources in relative to man the very citizens. Since people are endowed, government policies must be geared towards capacity development and human resources development. Let us look at, in few lines, stated below, how politics and democracy came to enhance the creative entrepreneurial potentials in the United States.

1. **ENERGY:**
   (i) The state government can and should play a crucial role… in energy future, according to people who are heavily involved in the energy industry or in the legislation, regulation or research affecting it.
   (ii) The most significant problem I see is our increasing dependence on unreliable sources of oil. All our problems stem from that.
   (iii) The Commissioner of Energy said he felt that the oil future looked bleak
   (iv) Conservation and solar energy… should be encouraged.
   (v) However… Hawaii, unlike New Jersey, had decided to marshal its natural resources sun, wind, and ocean power to try to achieve energy self-sufficiency. Adapted from AMERICAIS…. Pg 737; Recognizing Trends

2. **TECHNOLOGY**

   One thing that has helped change the American economy is technology the application of ideas, methods, and tools to the production of goods. Technology has helped Americans make more goods with less work. It has also helped Americans raise their standard of living, and it has given them more leisure time.

These and many more testify to the impacts of democracy to the creative entrepreneur.

Others are music, literature, transportation, painting, religion and most importantly, education. Once recorded is the history of how these creative entrepreneurs began to control the ships; harbours, railway lines, air transportation and even funded unclear power use in farming, oil exploration, space sciences; they developed strong business stratagiems to fight and eliminate unemployment, often times, with little government support; political reform systems, foreign trade and bilateral policies, currency and financial polices bothering on proper evaluation in effect to standards and costs of living relatively, transparency and accountability, prison reforms, environmental protection, law and enforcement, human rights issues, state policing, general security, defense and weaponry; the promotion of history and national heritage through the building of special public utilities such as libraries, museums, strdia, health centres and equipage; electricity for private and public use; industrial revolution, communications and satellites productions, newspapers, radio and television, care of the old, disabled and less priviledged. All these were available because creative entrepreneurship was valued and so human life had become too precious to be lost or wasted in any form.

CHAPTER 39

DECISION

The concept of decision takes its bearing from the simple right as endowed by the creator to make choices, while being aware and willing to go by the consequences of his own choices.

The choices a creative entrepreneur will make shall go far in determining to what extent he will make an input into the society. This will reflect in what to produce, for whom to produce, how to produce, his message, his market etc.

In order to make a decision, the following concepts stated below must be considered.

i. **Trends**: This implies that the creative entrepreneur needs to study the obtainable pattern and act for or against it. Trends do promote new ways and styles of doing things in such a way as to eliminate undue labor, wastage of time and resources and to promote beauty.
ii. **Abundance**: How to produce enough food, clothing and shelter for people to survive has been a problem for much of history. Changes in the ways that goods were produced improved some conditions. This, however, has an influence on what decision the creative entrepreneur must be involved in order to produce the best possible result.

CHAPTER 40

EXPANSION

The creative entrepreneur is one who would always seek methods of expansion. He needs to raise his stake, his importance and reputation by his services. In order to achieve this, here are some guidelines;

i. He must be willing to start where he is.
ii. In addition to contributions and possible investment, he must work smart and hard enough to link his various markets. Here, he must take the lead.
iii. He must make use of the various means of advertisement e.g. handbills, posters, internet, radio, television newspapers etc.
iv. He must take responsibility as a salesman; learning its ethics and procedures.
v. He must be willing to enlarge his capacities the fullest possible outcomes.
vi. He must educate himself through regular attendance of seminars, workshops and conferences. He must also be a reader.
vii. He must have an open mindset for greatness; for only expansion brings success and thereby greatness.
viii. He must be willing to accept and define the challenges which come from success.

•

CHAPTER 41

POWER

A certain feeling of power comes with being an entrepreneur. Power has been defined as the ability to influence others to think or act in a certain manner.

The creative entrepreneur must use his power and office of authority as a veritably tool for development and promotion of peace. His resources must be geared towards alleviating the sufferings of humanity

While operating sustainably, he must allow for teamwork. Team work promotes:

a) A sense of belonging among staff and people.
b) The flow of reasonable ideas.
c) The easy understanding of objectives.
d) Evaluation of set goals, achievements and records.
e) Trust and reliability among the organization's staff.

The wrong application of power may manifest in;

a) Lack of respect superiors and subordinates.
b) Loss of goals and objectives.
c) Poor communication flow among staff.
d) Loss of security.
e) Corruption.
f) Opportunists rather than developmentalists i.e. "Opportunity" here portrayed in the negative sense.
g) Lack of commitment to work and duty.

CHAPTER 42

PLANNING

Planning, I shall define as the systematic arrangement of future expectations based on foresight. He who fails to plan, plans to fail. Such a person assumes falsely that things and situation will automatically favour him. Little does he realize that things do not work that way. The creative entrepreneur will put the following factors into his planning sheet:

1. Purpose of product e.g. umbrellas for rainy season.
2. Budgeting: income and expenditures, sources of finance
3. Staffing and training
4. Time use
5. Systems development
6. Competitions
7. Packing and product cycle
8. Expansion and growth

Planning is basically divided into three: namely;

(a) Short term e.g. between 12 24 months
(b) Medium term e.g. between 2yrs 5yrs
(c) Long term e.g. between 5yrs 50years

Planning is a continuous process based on evaluative means and constant review. It is required that adequate information is obtained and recognized during the planning process.

SUMMARY

1. Opportunity is a favourable or advantageous or combination of circumstances. A creative entrepreneur must search for every opportunity to make a great difference.
2. Freedom is the condition of being free. Creative entrepreneurs work to be free financially and otherwise. Yet, there are responsibilities behind every freedom
3. Democracy, for the creative entrepreneur exists in two forms
   (a) Democracy as a form of government which favours creative entrepreneurship
   (b) Democracy within him to expand his creative abilities.
4. Decision is an inherent ability to make choices. It is influenced by factors such as trends and abundance.
5. Expansion is the ability to cover more grounds, taking other factors into consideration. The creative entrepreneur seeks opportunities for expansion always.
6. Power is the ability to influence others. Often the creative entrepreneur enjoys this facility. The extent of his power would depend on the impacts of his work on people.
7. Planning is the systematic arrangement of future expectations based on foresight. The creative entrepreneur must be a good planner
8. Good planning is divided into three namely:
   (a) Short term e.g. within 12 24 months
   (b) Medium term e.g. between 2 years 5 years
   (c) Long term e.g. between 5 years 50 years
9. Planning is a continuous process for necessary adjustments.

CHAPTER 43

BRANDING & DESIGN

Aims / Objectives

1. To teach the importance of bearing a name or keeping to a style for recognition.
2. To teach the steps to defining your own brand in making for excellence

BRANDING & DESIGN

In creativity, reality often appears as things. These things, objects and materials are not greater than ideas because they are substances of ideas. However, every idea being brought to reality is subject to being sold or marketed for profitability. For instance, what good is a book written without demand; or a school without students or music in a society of little or no appreciation for it. It is therefore necessary to define our work upon the steps of marketing and efficient sales. The process of meeting this requirement is referred to as the marketing forces. Certain factors are common to creative works of all spheres, whether Engineering, Arts, Communications, general sciences. The product must bear a brand and as well meet certain specified design. These would include;

i. Title strength
ii. Attractiveness
iii. Message e.g. Moral, Philosophical, religious etc
iv. Influence on society e.g. positive or negative
v. Colour maturity
vi. Pictorial clarity
vii. Catchy introduction
viii. Logos and imprints

DEFINING YOUR OWN BRAND

To simplify these principles, we shall adopt these stated in the EQIUNEWS MAGAZINE published by ETB;

1. **The Principle of specialization:**

A great personal brand must be precise, concentrated on a single core strength, talent or achievement. You can specialize in one of many ways; ability, behaviour, life style, mission, product, profession or service.

2. **The Principle of Leadership:**

Endowing a personal brand with authority and credibility demands that the source be perceived as a leader of the people in his / her domain or sphere of influence. Leadership stems from excellence, position or recognition.

3. **The Principle of Personality:**

A great personal brand must be built on a foundation of the source's true personality, flaws and all. It is a law that removes some of the pressure laid on by the law of leadership; you've got to be good, but you don't have to be perfect.

4. **The Principle of Distinctiveness:**

An effective personal brand needs to be expressed in a way that is different from the competition. Many marketers contract middle of the road brands so as not to offend anyone. This is a route to failure because their brands will remain anonymous among the multitudes.

5. **The Principle of Visibility:**

To be successful, a personal brand must be seen over and over again, until it imprints itself on the consciousness of its domain or sphere of influence. Visibility creates the presumption of quality. People assume because they see a person all the time, he must be superior to others offering the same product or service.

6. **The Principle of Unity:**

The private person behind a personal brand must adhere to the moral and behavioural code set down by that brand. Private conduct must mirror the public brand.

7. **The Principle of Persistence:**

Any personal brand takes time to grow and while you can accelerate the process, you cannot replace it with advertising or public relations. Do not change your personal brand; be unwavering and be patient.

8. **The Principle of Goodwill:**

A personal brand will produce better results and endure longer if the person behind it is perceived in a positive way. He / she must be associated with a value or idea that is recognized universally as worthwhile.

Remember, building a brand takes time and is not based only on what your say, but also on what you do. It is much easier to maintain a personal brand based on your true authentic self than it is to build one on an artificially created persona.

SUMMARY

1. The process of meeting requirements for success is termed marketing forces.
2. To define your own brand, one may adopt the following principles:
    (a) The principle of specialization defining precision, based on ability, behaviour, life-style, mission, product, profession or service.
    (b) The principle of leadership steming from excellence, position and recognition in a sphere of influence.
    (c) The principle of personality based on the extent of personal development and appreciation of others.
    (d) The principle of distinctiveness making your difference clear from the competition
    (e) The principle of visibility creating the presumption of quality.
    (f) The principle of Unity adherence to the moral and behavioural codes and standards set down by both private and public regulations.
    (g) The principle of persistence allowing for the process of growth and development of such brand.
    (h) The principle of goodwill creating the authentic perception of a good idea and value recognized universally.
3. It is much easier and better to be original in branding than developing an artificiality of another.

CREATIVITY RESOURCES TABLE

Aims / Objectives:

1. To teach the student the kinds of resources available.
2. To teach the proper deployment of resources available for the best management
3. To discuss the potentials of the brain to develop usefully every resource gainfully.

CREATIVITY RESOURCES TABLE

Every form of activity, whether found in business, invention, education or administration cannot function without the employment of certain resources. This also applies to the world of creativity. Briefly, we shall study what I term the creativity resources table.

Resources are those necessary ingredients which when brought together combinatively promote the accomplishment of an anticipated activity. There are inherently five kinds of resources:

1. **Natural resources**: which are God-made. These include minerals e.g. gold, tin, bauxite, crude oil etc
2. **Artificial or Man-made Resources**: which eventually form the basic raw materials for the development of another material e.g. caustic soda for making soap etc
3. **Human Resources**: This is by far the most valuable of all resources. It deals with human employment in order to accomplish certain objectives, often profitably.
4. **Essential Resources:** Original in thought in relation to this field of study, these include Time, information, access and expressiveness. Without the proper employment of these resources, either combinatively or single handedly, this set objectives may not pass finally for use by the customer. Implicatively, the process of production is not complete except it reaches the final consumer.
5. **Capital Resources:** This is the raising, accounting and investment process of money or cash. It is the management of finance.

In simple terms, we shall describe the creativity Resources as ETIM, CP. TM; grouping them; we have:

Groups	**A**	**B**	**C**
	ETIM	**CP**	**TM**
Resources	E Energy		
	T Time	C Capital	T Tools &
	I Information	P People	M Machines
	M Materials		

In a shorter code, the table above may be known as the Creativity Resources Table. In order for memory, we may quote it as thus:
"**ETIM** Construct People's Time & Money". (Etim being a name in the South-Eastern region of Nigeria Crose Rivers State, precisely).

The above resources must be duly considered by analysis before one finally commits himself to selling into any venture. It may also be noticed that by the resources table, all genres of study, sciences (TM) (E,M); Arts (I,T) & Commercial / Administration (C,P) are combinative. It proves that creativity cuts across every sphere of human activity. No one is limited in any form. The careful employment of the thinking faculty would be supported by the question posed by Dr. Whitt N. Schultz of Kenilworth, Illinois, who states:

"Are you using your brains? You have billions of brain cells. In fact, your brain has the remarkable capacity to take in, process, program and utilize more than 600 memories per second for 75 years (and more!). That's 51,840,000 bits of intelligence per day that your mental computer can handle!. Are you using your brain and the brains all around you?

You and every other normal person have four brain powers; they are:
Your absorptive brain power That's the power to take in knowledge through the "gateways to your mind", your senses. Open up your mind. Let the sunshine in!
Your retentive power This is your memory. Stored in your "library of knowledge" is every thing you have experienced in your life time, up until now.

Your judgemental power - Your power to judge, to make choices, to respond to situations based on your facts (which are often incomplete). This power is usually so active that it gets in the way of...
Your imaginative power This is the tremendous strength of your imagination. As youngsters, our imaginations were alert and utilized constantly. But unfortunately, throughout the years, our imaginative powers were stifled by what I call "killer phrases" "it won't work"... "It's not in the budget"... "we tried that before and it doesn't work"! "No way", etc.

It's the great growing and mostly unused power of the imagination which most business people fail to generate. And yet today, if ever there is a time to use our imaginations to come up with solutions, that time is now. One way to generate renewed mind power is to bury forever the Lawrence J. Peter "Peter Principle". You know that theory "In a hierarchy every employee tends to rise to his level of incompetence".

Therefore, a proper estimation of these resources would inevitably bring a person to the point of achievement in whatever goals he sets to achieve.

SUMMARY

1. Resources are those necessary ingredients present which when brought together combinatively promote the accomplishment of an anticipated activity.

2. There are five kinds of resources:
   (A) Natural resources i.e God-made e.g. gold, tin, bauxite.
   (B) Artificial or man-made resources e.g. caustic soda, paper
   (C) Human resources; people available for labour & business.
   (D) Essential resources e.g. Time, information, access and expressiveness
   (E) Capital resources i.e. money for business purposes.
3. The resources table may be memorized as thus; "ETIM Construct People's Time & Money" Where E Energy; T Time; I Information; M Materials; C Capital; P People; T Tools & M Machines.
4. Every person has four brain powers:

1. Absorptive brain power the gateway to take in knowledge
2. Retentive power this is the memory known as the library of knowledge
3. Judgemental power the point of decision making
4. Imaginative power the power to create mental pictures in line with our desires.

CHAPTER 45

TALENTS MANAGEMENT

Aims / Objectives:

1. This study aims to inculcate into the student the techniques applicable to talent management.
2. To teach that the principles of management are universal and can be applied to talents as well.

TALENTS MANAGEMENT

Conventionally, the word talent means a natural or acquired ability, aptitude; a natural endowment or ability of a superior quality; a person with such ability; as well as any of various ancient units of weight and money; while management is the act, manner, or practice of managing; the person or persons who manage an organization; executive ability; but the word manage implies the authority to direct, control, or handle; to make submissive; to direct business affairs etc.

Combinatively, talents management may be defined as the ability or capacity to manage or run the business affairs, and any other, as it concerns talents. Such a person would be a talents manager or skills resources and development expert with special knowledge on careers management. This entails part of the responsibilities of a creative scientist.

Intuitively, the word talent as defined above refers to three important aspects of our common existence namely natural ability or endowment; some sort of superior quality and units or weights of money. This implies that certain characteristics are similar to the three.

Characteristics of Talents, Quality & Money

1. They are often endowed or given for common use.
2. They are valuable means of exchange to satisfy needs and offer services.
3. They exert certain sense of influence when used or mentioned.
4. They are measurable
5. They are quantifiable
6. They are widely acceptable
7. They have a common source which is ideas.

From the above features or stated characteristics, it proves that ideas is the central factor to their operations. Talents are developed into objects, or needed substances of high quality and perhaps sold for money but all depending on the superior level of quality or strength of the idea. A good idea, in order to reach its substantial peak must be properly guided and developed according to certain guidelines or operational methods. These guidelines are what may be referred to as talents management. The simple guidelines below are expected to help you manage your ideas and talents to yield expected results.

Guidelines to Talents Management:

1. Statements / Records

Psychology has proved that over 10,000 ideas pass through our minds daily. Where the problem of this statement lies is in the fact that people do not often take them serious enough to jot them down somewhere. Personal research proves that initial ideas often occur to our subconscious as simple specks. With time and combining other necessary factors, they begin to make certain sense and then we become inspired. Sometimes, this idea becomes a spare part to the other. It is therefore, necessary to make statements, of our ideas once we receive them. Such statements would include details such as the title of the idea; time and date of occurrence, existing situation, place, purpose and potential capacity (market); requirements and strength of longevity, other details.

Your Ideas Diary would look like this;

STATEMENTS OF TALENTS / IDEAS

1. Title of idea: ______________________________
2. Time & date of occurrence: ______________________________
3. Existing situation: ______________________________
4. Place of occurrence: ______________________________
5. Perceived / Initial purpose of idea: ______________________________
6. Potential market capacity / Audience (Age, class, season etc) ______________________________
7. Financial & Material requirements (Estimates only) ______________________________
8. Strength of longevity (years): ______________________________
9. Other details: ______________________________

2. Quantitization of Ideas

After a considerable period of making various statements and records of ideas, then personal or organization selection may be made through the grouping or specific method. This is what is termed the quantitization of ideas. Sometimes, certain ideas tend to match when combined and so are selected to meet a particular purpose. This is known as grouping. When a specific is singly selected on its merit, this is referred to as specific selection. Ideas will often be selected on the basis of easier implementability, and favourable financial and situational opportunities.

3. Rights

After selection and development and experimental market surveys, then it is time to obtain your legal rights to such an idea or product. For literary works, musical and sound recordings, copyright is applied for, and given to the idea owner, often by the Copyright Commission of your country or authorized agency for such purpose. In the case of products or inventions to be produced industrially, what is obtained is known as Patent Right. This is often obtained from the chamber of commerce of your country or any authorized organization to do such. If the idea is large enough you may need to register a simple business name (Enterprise) or incorporate a limited liability company. This is where you have become self employed. Records have it that Thomas Edison had more than 1,000 patent rights to his credit. These rights are powerful because they individually grant you the rights to the everlasting proceeds or wealth accruing from it and if well managed and protected, would become assets to the next generation. However, your ideas must be protected from intrusion or theft. It must be guarded closely. After copyrighting, the following steps may be made for expansion purposes.

1. Selling: to organizations in order to make money. Often carried out as Services.
2. Partnerships: to build a relationship with a view to create business relations. The essence of business partnerships is to create a merger, or large outfit to meet with the requirements of a large market and make more money. Royalties is payable to the owner of the copyright as negotiated. Depending on the nature of the product or service, partnerships could hold between the copyright and organizations such as Non-governmental organizations (NGO's) private companies, government agencies and humanitarian organizations. Expansion programs would include, advert placements, seminars, interviews, press conferences and lectures etc. Then succession plans to help for control and proper trust to your contributors must not be neglected. It also gives everyone you wish to be part of your inherence a sense of belonging.

SUMMARY

1. Talents, quality and money share the same qualities and characteristics.
2. Guidelines to talents management are:
   i. Making of statements and records

   ii. Quantitization of ideas
   iii. Rights management
3. Rights management include obtainance of:
   a. Copyrights; for literary works, musical and sound recording
   b. Patent rights; for industrial works and inventions
4. Rights benefits include:
   i. Selling
   ii. Partnerships
   iii. Succession plans

CHAPTER 46

CURRICULUM DEVELOPMENT & CAREER CAPACITY OF THE CREATIVE SCIENTIST.

Aims / Objectives:

1. To teach the student to understand the nature of the principles of creative sciences.
2. To discuss the various professional and ethical responsibilities of the creative scientist.
3. To introduce the student to the guiding body and philosophy of the creative scientist.
4. To differentiate the general study importance of creativity from the professionalism attached to the field of creative sciences.

CURRICULUM DEVELOPMENT & CAREER CAPACITY OF THE CREATIVE SCIENTIST.

1. Nature of study: The study of the field of creative sciences covers distinctively four primary aspects of interest;

a. Psychological: This studies the workings of the mind of a person or natural endowment and character building attributes of a personality. It studies the influence of our reactions in solving our problems individually and collectively.

b. Spiritual: Here, the studies focuses on the recognition of a supreme creator whose image is directly inbuilt into man; and has the capacity to control man by a certain presence of internal power (conscience). It supports the principle of creationism and the P. M. Aku's philosophy of paralleled opportunistic creation.

The philosophy of paralleled-opportunistic creation (POC) states that "When God created the universe, he left certain unfinished works as opportunities for man's creative refinement in every area of creation".

Proving this, it may be discovered that electricity and electrons (man-made) draws their origin from lightning which occur as positive (+) and negative (-) ions in the atmosphere, the sky was made ready for nutrients recycle; birds for flight sample.

In order to understand these further, let us see the few in the table of POC table below.

PARALLELED OPPORTUNISTIC CREATION TABLE (POC)

S/n	God's Creation	Man's Opportunities
	1. Sky	1. Centre of scientific study of recycle.
	2. Water	2. Clouds & weather elements for agriculture.
		3. Flights of birds for aviation and aerodynamics.
		4. Use of radioactive signal to transmit effective communication e.g. TV & radio signals, satellites etc
		5. Lightning spark / electrons form basis for electricity.
		1. Acquatic studies
		2. Fisheries
		3. Shipping: commercial and technological
		4. Water necessary for living
		5. Importation and exportation of goods
	3. Earth	Plane

Forests for hunting, forest resources cash crops e.g. timber for housing, fruits and vegetables / agriculture for living; study opportunities; conservation; land for accommodation; industrialization and veritable uses.
Minerals exploitation e.g. crude oil, gold, tin etc. metallurgy, geology etc.

The spiritual aspect of this study also discusses the influence of muses, and the proper realization of special intentive powers such as vision, imagination and interest.

c. Practical and social: The aspect of the study which deals with direct responsibility guiding every service rendered profitably or unprofitably to humanity. This is based on choice and special abilities associated with such practice.

- Scientific analysis: formulae and laws

Based on these, the field of study of creative sciences may be described as a psycho-social science (Advanced study) or social science (intermediate study Secondary education). This study has the sole purpose to impact humanity through resourcefulness and service.

2. Creativity Combination Principles I, II. & III

The P. M. Aku's creativity combination principles are stated below:

Principle 1: No field of endeavour exists without an aspect of creative development as long as man is involved.

Principle 2: The strength of every field of services is determined by the extent of creativity applied to it.

Principle 3: Every form of human service requires creativity in order to remain effective and significant in the face of competition.

The implications of these principles would require a stage in service development such as:

a. The combinative study of creative sciences and another field e.g. study of accountancy and creative sciences would mean that such a person is a creative accountant or an accountant and a creative scientist; or a creative banker, a creative pilot; a creative architect; a creative writer; a creative footballer etc.

The major responsibilities of these specialists ranges from teaching (education); strategists (administration, military, security); players in departments of customer/ client transactions, research & development; policing (major crime detections and investigators); departments of intelligence and logic in various sectors; and often play critical roles to solving problems by simple understanding yet complex operations in politics and governance as well as inventions; depending on grade of study and combined field.

b. The professionalization of the single study of creative sciences, in order to further develop one's core personal strength e.g. writing, drawing, oratory etc.

3. <u>Educational Qualifications, Equivalents & Office of Responsibility of the Creative Scientist.</u>

The basic study awards and qualifications in the field of creative sciences is structurally divided into two stages:

i. Basic Diploma

ii. Professional

Below is a summary of these

AWARD CONVENTIONAL EQUIVALENT OFFICE / CAPACITY OF SERVICE

1. Basic Diploma (BD) Equivalent to Ordinary National Diploma (OND) Works in primary schools as a creative science teacher Basic 5 & 6 or 4 & 5; as applicable; or other simpler operation in a business e.g. receptionist etc.

2. Professional Stage

i. Male Master of Creative Sciences M. Cs (Hons)

Female Mistress of Creative Sciences Ms.Cs (Hons) Equivalent to the universities 1st degree with a honours.

A creator of a copyrighted work.

Self-determined Education secondary school teacher and other operations.

May be self employed.

ii. Male Genius of Creative Sciences Respectably G.Cs (R)

Female Mistress De Genius of Creative Sciences Respectably MDG. Cs (R) Equivalent to the advanced stage of conventional masters.

Owner and facilitator of five patented works considered outstanding. Self determined. Education Lectures the creative science course at the ICMs; colleges of education; special higher institutes with proper approval from the monitoring and establishing organization.

iii. Male Sage of creative sciences for life. S.,Cs (L)

Female Mistress De sage of creative sciences for Life MDS. Cs (L) Equivalent to the most respected stage of the conventional PHD. Owner and facilitator of a minimum but not limited to ten patented works considered outstanding. If up to 15 works, inducted into the hall of creators and becomes a chartered sage. Self determined. Education acts as a consultant; runs a branch; operates a private centre after due approval by the ICCS to prepare special exams for applicants of the G.,CS (R); and S.,CS as applied male or female annually.

1. Pioneering organizations of creative sciences.

1. The Institute of creativity and management studies (ICMS)

The ICMS began in September 6, 2010 as an establishment in Igando area of Alimosho Local Government Area in Lagos State, Nigeria. It was founded by Peter Matthews Akukalia the first creative scientist ever to develop the field. He is by standards the first chartered sage of creative sciences.

Role of the ICMS

Its major objective is to train a new breed of professionals known as creative scientists. It also trains the student from Basic Diploma stage to the Master stage and then prepares them for professional examinations on this field.

2. The Institute of Chartered Creative Scientists

Purpose: Set up to give due recognition to those who are deserving of such recognitions as necessary.

Administration: The ICCS is managed by the chartered and headed by the President / CEO of the UTMOST PEAK RESOURCES; as well as assisted by the board of Governors. The Board of governors are those inducted into the Hall of creators, who stand in recognition above the sages. The ICCS sets examinations and marks them. By approval and certain fees, the ICCS approves centres for sages to run institutes as well as approval of text kits. Its presidency is hereditary, yet democratically run in advance to the President / CEO but advised by the Board of Governors. This is to avoid power struggle.

SUMMARY

1. The principles of creative sciences may be termed a psycho-social science because it discusses;
   a. Psychology: workings of the mind;
   b. Spiritual: the recognition of God in creation and the strength of conscience;
   c. Practical and social: guide to relationships and results, responsibilities to the economy.
   d. Scientific analysis: laws and formulae, here, the renaissance study of Pmakumatics is discussed
2. This study is guided by the P. M. Aku's creativity combination principles I, II, and III
3. Principle I: states that no field of endeavour exists without an aspect of creative development as long as man is involved.
4. Principle II: states that the strength of every human service is determined by the extent of creativity applied to it
5. Principle III: states that every form of human service requires creativity in order to remain effective and significant in the face of competition.
6. The level of qualifications of a creative scientist determines his office of responsibility. These academic qualifications are: Basic Diploma (B.D); Master / Mistress De Genius of Creative Sciences (Respectably) G,CS (R) / MDG,CS (R); Sage / Mistress De Sage of creative Sciences (for Life) S,CS (L) / MDS,CS (L) and Chartered Sage Ch.S
7. The organizations which administer over the field of creative sciences are:
   a. **Institute/Department of Creativity and Management Studies:** Which train and graduate creative scientists professionally as a part of the parent institute.
   b. **Institute of Chartered Creative Scientists:** Which responsibility is to induct chartered creative scientists into the society and monitoring various guidelines for professionalism in the field of practice.
   c. **Hall of Creators**: For recognition of creators who have reached the chartered stage or level.

d. **Ministry of Creativity:** A national resource for endorsing the creative organizations e.g. Private museums etc; funding certain works of creativity after recommendations from the institute of chartered creative scientists; approval of the various professional centres after endorsement by the institute of chartered creative scientists and hall of creators. Through this ministry, certain difficult national problems from other sectors are sent in a comprehensive report for brainstorming by the ICCS and Hall of creators/ and then recommendations. Also, the ministry of creativity sees to the posting of creative scientists as strategists to every sector of the economy; minimum of 10 persons in a department.

Professional Distinction In Creative Studies.

There is the ever present need to discus the professional distinction between the General study course of this program and the professional study course. This is most understood in a table as stated below:

S/N	General Study Course	Professional Study Course
1.	Perceived as a necessity in every university and tertiary institutions	Is a personal choice after higher learning in addition to a specified study or discipline.
2.	Introduces the person/ student to the basic knowledge of creativity while still in school	Prepares the student to become an economic strategist professionally, to work in line with his calling and training.
3.	Student receives no special award for his knowledge in creative principles	Student becomes awarded at each stage of his professional programs
4.	Breeds general student based on conventional fields	Breeds creative scientists in addition to conventional study.
5.	Most awards are based on studies	Further awards after preliminary awards are based strictly on works.

6.	Training is generalized and done by creative scientists who by training are teachers or lecturers	Training is specific and managed by professional creative scientists and the various professional organizations e.g. ICCS, Ministry of creativity (MOC); Hall of creators etc.
7.	Conventional study may train students to be job seekers	Strategic study trains students to become job creators and employers.
8.	Conventional study may be in respect to other fields e.g. law, medicine etc	Strategic study views every conceivable field available and understandable to the mind of the creative scientist.
9.	Conventional study may promote shift of responsibility to the government and economy.	Strategy in creativity makes you responsible for existing problems in your field. The creative scientist, irrespective of the area of service, has an obligation to take personal responsibilities to solve problems, or at least profer solutions.
10.	Strengths in study is drawn from conventional thinking	Strengths, here, is drawn from the creative mind
11.	Conventional thinking works by rationalism	Creative scientists work by logical ideas
12.	The philosophies of realism or idealism is promoted	The philosophy of raydealism is inherited.
13.	Conventional study may produce mere thinkers and speculators	Professionalism produces strategists, innovators and result oriented persons.
14.	General study promotes its own in addition to laid down rules and regulations	Strategic study breeds leaders and workers of the silent Empire the emerging creative economy.

Relationship Between Creativity and Careers

The basic relationship which exists between creativity and careers or professions is strategy. Strategy is the careful planning and management involved in every field of human endeavour. Such a person who strategises is called a strategist. The basic essence of strategy is to introduce innovations, change, new style, analysis, better understanding of an operational system and growth in income. Strategy develops new ways of accomplishment in less energy and time, often used to accomplish such work.

A careful study of American historians in the Great Depression proved that a new awareness of the power of style borne out of strategy improved their lifestyles from the 1920's to the present date. The book America Is States; "During the 1920's, Americans developed new styles of life. New inventions freed people from certain tasks. The automobile allowed them to go places more often. Vacuum cleaners, canned food, electric refrigerators, and gas ovens cut down on the time needed to do house work. Many women found they had more free time and greater freedom. They worked outside the home, studied at and graduated from college, and entered professions".

There is actually the fact of creative living, creative working and creative professionalism.

CHAPTER 47

CREATIVITY IN PROFESSIONS

- Scope of creativity
- Professional Applicability

SCOPE OF CREATIVITY

It is important to understand the fact that every career is a function of human endeavours. Human decisions and actions may lead either to a creation or a destruction. Jobs are in existence because certain courageous people brought into reality. In other words people create jobs, keep jobs and sustain jobs in order to better human life and interests. This is the proof of the strength of creative endowments and abilities placed upon each person. If humans are not able to properly sustain themselves, then it is a proof of the poor use or inefficiency of the powers imbued upon her from creation.

The scope of creativity stretches beyond the very universe itself. Creativity is the first subject of universality. It is the core strength of existence. It is the linking thread of all activities. It is the proof of living realities. Creativity is the search of the sciences, the beauty and expression of the Arts, and the guiding management of the commercial wing, the thoughts of the philosopher and the lessons of the teacher. Every field is a body of creativity and only those who apply creative techniques to their career or activities will reach the top and stay there as well.

Let us examine the relationship between the study of creativity and the various careers. In order to understand this better, study the following statements.

i. In every profession, competition exists and only creative people excel.
ii. There is no field of human endeavour without an aspect of strategy and creativity in it.
iii. Customers, as every other people, wish to see new, inventive and exciting products and systems of performing tasks.
iv. Innovations colour the brown earth and make skyscrapers from sand and mud.
v. People who refuse and resist change surely do risk a future of bliss.
vi. There's no end to creative abilities because there is no end to God's abilities
vii. Birth signifies a new system in generational development, while death is a transition to the next level
viii. The first evil ever repugnant to functionality of a body, mind or system is poverty because it represents dryness, scarcity, lack, inactivity, instability, little or no strategy and if unchecked, eventual death.
ix. Creativity implies tapping from the greater resources to run life's engines
x. There's no problem without a solution.

The statements above indicate, that every field of study and work ought to have creative scientists or creative strategists within their system. In order to emphasize this point, I shall propose the theory of a creative economy.

P. M. Aku's theory of creative economy states that the strength of the society is wrongly measured by its wealth but rightly by the positive impact of each individual contributory to such a society.

The careful study of every field shows that people are in charge, thereby there is a need to comprehensively differentiate, for instance, an accountant from a creative accountant; or a pilot from a creative pilot. Every strategist anticipates danger but a creative strategist foresees danger and plans ahead to avert or control this danger.

A creative scientist or strategist is a person who majors in creative sciences, while its addition to another field differentiates him e.g. a creative preacher, or a preacher and creative scientist. It is however, necessary to underscore the need to recognize such a person as one who is certified in this field.

PROFESSIONAL APPLICABILITY OF CREATIVITY

The first essence of professional applicability of creativity recognizes the essence of strict adherence to its basic principles such as research, interest, inspiration, imagination and environment, then the understanding of basic human attitudes in creative philosophies.

It is thereby, highly necessary for us to examine few areas where creativity could keep us afloat the waters of time.

1. Education:

The education sector, in any society, must realize its place in national development, mind building and leadership training. The creative educator or teacher should do more personal work, understanding the specifics of his class, asking interesting, practical questions which pertain towards problem solving skills of our day; applying sensible humour when necessary, perhaps to vibrate a seemingly dull class; not disobeying conventional practice but stimulating the mind of each pupil or student. Creativity is liberal and so must the mind which taps from it, since every living person has a measure of it. It's time the education sector, in my honest appraisal re-engineer the syllabus/ curriculum to include history and biography studies (for human appreciation, integration and national unity); geography (to appreciate culture and diversity); agriculture, applicable youths finance education and better teaching standards; such as class teacher and students discussion hour where under supervision, issues bothering on unemployment, skills development, career challenges and prospects and morality is discussed; taking into cognizance students' questions and worry to be booked and researched for development.

2. Religion

Here, rather than more screams from the preacher's platform or pulpit, a better strategy would be teaching the practicality of the scriptures based on research, analysis and real understanding; encouragement for members to study, upgrade their own knowledge through personal study and leading a more peaceful life, in reverence to God, the Almighty creator. Such one is called professionally, a creative preacher.

3. Business:

The best kind of business is one whose impact makes for ease from a drudgeous way of life. Sincerity and honesty makes the best business. Strategies should be developed to increase solution products to solve problems and needs at affordable rates.

The creative business person seeks nothing but the satisfaction of his customer or client.

4. Computer & Information Technology:

Perhaps, this sector has received the best of creative recovery in the recent years. Simply put the creative options chose since the discovery of the computer, its construction, its various uses as necessitated by the need for time and information management has made the computer a friend next door.

Interestingly, the first computers (the Abacus) was first conceived and used by King Solomon of the Bible. Like every other sector, its impact was not known until the creative information giants Microsoft corporation, Dell and others saw through their creative eyes. Bill Gates is no new story.

Further, its importance today on the creative terms streams from software programming (e.g. Excel & Dbase), ATM's and E-banking, Desktop applications and Graphics, Power Point etc. Imagine applying creativity to solving other problems of our various needs, just as much as there's been in this sector. Every thing needs creativity.

5. Security & Defence:

Beyond mere policing or defence, there should be a department of strategic operations which have therein creative scientists with combined background study of crime detection, law enforcement and combat of terrorism. The first criteria of a creative scientist or strategist is to handle the situation based on its own merits, while considering the best answers from his practical analysis.

6. Engineering:

Collectively, I shall refer to power generation, scientific inventions, production and industrial machines, etc. There's need to create such departments of strategy in order to experiment new and better ways of performing the same task.

7. Government & Politics:

A creative politician ought to have devised, tested and proven a certain style of governance ever before contesting. Such politicians ought not make vain promises dependent on irrational reasoning; rather they should be ready achievers; not enemies but analysts; not luck Sayers but record- tested; perhaps with guarantors whose image are at stake if such, political office holders fail by provable corruption or guilty of any crime against humanity; willing to lead not rule; not looters. The creative political holder must work by himself; and with others who are strategists.

It is necessary to realize the need for creative lawyers, creative bankers, creative musicians, creative writers etc. The basic prerequisite is originality from your treston. By this, we shall do things anew, and bring back the days of honest work, deserved pay and better living, then we shall love what we do and do what we love.

CHAPTER 48

UNIVERSAL RELATIONSHIPS

In life, the first understanding of nature and creation is the existence of variety. There is variety in variety itself. Various aspects of creation exist in phases and levels. These phases and levels are determined strictly by their original purpose and systems of functionality. Here, we may notice for instance the existence of oceans and their wave strengths, rivers, oasis and swamps in relation to the marine, jungles, forests, bushes and simple flowers in relation to plant life; the planets when judged from their sizes, and distance from the sun, the stars, man's creations of technology, man himself and even knowledge. Among all these exists likeness within a group; and simultaneous distinction within the same group. Yet, they must all work together, each being an important factor of the other, in order to achieve a common purpose. What binds them all is what I term "universal relationship".

Creative relationship exposes our minds to the importance of teamwork, trust, reliance and communication. Here I shall define what I term the law of universal relationship.

"The law of universal relationships states that functional systems, though distinction uniqueness, works through a series of relationships to achieve a useful purpose and meaning".

This law proves that an economy would function based on the following grounds;

1. That each citizen understands himself and mind to achieve a useful purpose.
2. That each person appreciates his own existence and strives positively to make it better.

3. That the existence of unfavourable conditions is negated by evil tendencies which must be stopped by constant orientation.
4. That opportunities abound when it is harnessed firstly, from the mind and then outwardly.
5. The strength of relationships built is also contributory to the strength of excellence achieved, yet each person's contribution is exceptional to the overall good of the whole.

The importance of the law of universal relationships works upon the laws of a creative economy. When the mind fails, the citizens fail and thereby the nation fails yet the nation plays a fundamental role of systematic functionality. Nevertheless, it is needful to build useful relationships in order to foster excellence in all that one strives to achieve.

Universal Web of the Creative Economy

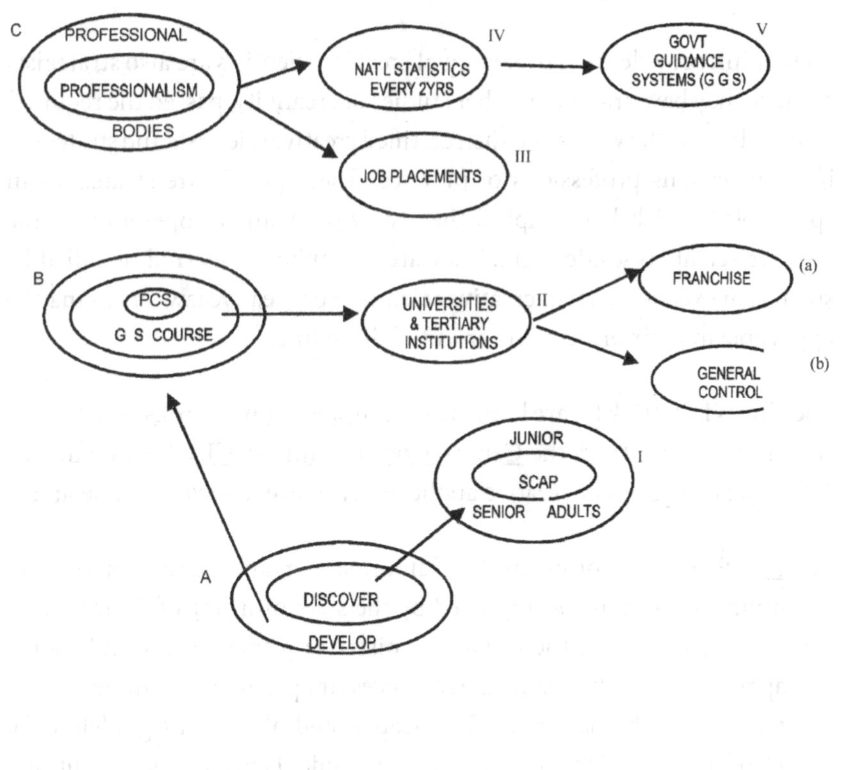

Explanations of the Universal Web of the creative Economy

The circles A, B and C are known as the Central Cosmos Links (CCL). They define the projected sources meant to usher in the creative age. Here, A seeks to discover and develop the potential in the person through the B link. The B Link contains the Knowledge Impartation Procedure (KIP). The KIP empowers the PCS to become the authoritative source of creative studies as a General Studies (G.S) course for universities and tertiary Institutions including the Public Federal, States, Local (e.g. vocations schools); colleges of Technology, Colleges of Education and private all over the federation.

The B Link qualifies a graduate of any academic discipline to train to become a creative scientist: The C Link empowers such training. This may be done, supervised and empowered by the professional bodies. These bodies include the ICCS; Hall of creators and the government assigned ministry of creativity.

The C Link provides that professional creative scientists are also strategists because they have trained in other studies of creativity, passed the required tests and proved by works and then certified creative scientists and strategists in their various professions of practice. These people are strategists in "principle", which here implies that every guideline of operation of the creative scientists guide them. They are not publicly referred or called by such term except the term creative scientists; or creative lawyers as may be applicable to their professions or initial discipline.

The circles I, II, III & IV are known as the minor Cosmos Links. They derive the in sustenance from the Central Cosmos Links (CCL). They define the following categories of creative applicability to every society. These are:

1. **SCAP**: Its categories are the junior, senior and adult. For the full utilization it must be approved by the state ministry of Education as compulsory; while the Federal Ministry or Department of Education approves its result / certificate as a necessary pre-requisite among others for entry to higher study. The Scap would also act as guideline for Department and faculty fixing of new students to tertiary institutions, as well as for already existing graduates and adults in general.

2. **GENERAL**: Others are referred to as the general. However, VI and V work together. The annual published statistics of the various persons and their sectors would help the recommendations of the organization to government on what industries or vocational programs to be developed at a time. This is known as the government guidance systems (G.G.S).

CHAPTER 49

PREDICTIONS OF A CREATIVE ECONOMY

Every study worth considering in the affairs of mankind has always had a profound effect. This is because people, guided by the philosophies of such study are expected to behave in a certain manner in response to the imparted study. The study of creativity is no exception. The essence of law is to ensure that there is proper settlement of disputes, justice given to him whom it is due, punishment metted out to offenders after being proved guilty thereby ensuring the reduction of chaos and anarchy in a society; the essence of philosophy is to pattern the mind of the people towards civilization, engineering, out of science and technology hopes to solve certain problems which men may suffer, using machines. Yet, it is pertinent to understand that the fathers of every field of study or profession, at one time or the other, nursed silent dreams or ever foresaw a certain breed of society ever before sweating it out. Of course, the religions and Human Rights Activist, Martin Luth or King Jnr. Once screamed, "I have a dream", and the rest of that story is history. No study or principle or philosophy or orientation in any form is complete without the predicable outcome of such in the society of people.

It is therefore, highly probable, and thereabout necessary to state and explain the predictable outcome of creativity on the society. These are my predictions and I hope they are to be assessed on the 5 year Assessment Basis. For history is a review of records either positively or negatively. These are the predictions of a creative economy.

PREDICTION 1: CERTIFICATION

I see an economy where every person, young or old, great or small, educated or uneducated is respected along by his / her certification; that is SCAP certified.

This is possible because the individual would in a time to come become relevant by the input contributed to the society. Here the students career analysis program considers everyone a student of life. It respects choice, interests and over all personality and produces results based on individual's records, graphs analysis, calculated level of endowment in each person, encourages wider scope of career choice which comprise Academic based and vocational; local and global based courses, research and innovativeness.

Since, there are students who might need to obtain work experience after tertiary education, their Psycho-analysis test (PAT), or Personality Attitude Chart (PAC) would be conducted helping to determine their level of competence, level of interest, and better understanding of such an individual.

The SCAP certificate would also ensure that employment is strictly according to calling and study ensuring fulfillment in the work place, avoidance of misplaced priority a situation where a person works differently in a job which has no bearing to his calling; people would also become more of self-employed.

TRAGEDY: The tragedy of this age is that graduates who ought to lead intellectual activity graduate hoping to find the job which they ought to be creating; while the experienced uneducated is not properly certified in a good vocational institute after learning a trade or vocation in order to prove his competence.

PREDICTION 2: EMPLOYMENT

I see an economy after this study, where citizens are re-orientated and they begin to dig deeper from their own reservoirs of creative abilities to create jobs, in so much that few people would likely work for others competently while the number of available employment would far outweigh the number of job seekers. By the PAC test, the curriculum vitae (CV) would be upgraded to include the SCAP and PAT as well as the interpretation by the relevant organization of the level of interest of this person in relation to the job sought.

PREDICTION 3: SECURITY

The best form of security is not total punishment but re-orientation of the individual's mind set and thinking. With the promotion and reward of creativity, citizens tend to understand that they ought to contribute to the society through hard work and resourcefulness. The Police would assess culprits by their SCAP records, holding parents and youths responsible for not using their minds to think up positive contributions. If there were no funds, for instance, to attain a higher education, then there would definitely be no excuse for not learning a recorded vocation or trade. Furthermore, a youth who begins life early, would definitely be able to plan for a higher education.

PREDICTION 4: STANDARD OF LIVING

Take for instance, a street which has in it, 5 seamstresses, 8 watch and clock makers, 10 hair dressing shops etc all professionalized and innovative in their style of work by the knowledge of this course; the greater chances are that a healthy competition would ensure better quality of products, lower cost, likely provision of local training and raw materials by importers at cheaper rates (if necessary) and a better living standard. Why? Because everybody wants to sell. Imagine a world where a wrist watch would cost as low as N10, and would be considered a waste of time to go repairing it rather than quickly buy a new one. That is the world I see, and this you must see also. Youths would be productive and ready to design new models to be produced and sold as their developed minds would readily conceive; no matter the field.

PREDICTION 5: EDUCATION

The SCAP and the Oracle expose courses and vocations of study as it is obtainable the world over, this knowledge would also lead to an array of choices, thereby would reducing the undue pressure presently obtainable, on the relatively few courses studied and vocations learnt. Every faculty has many branches of study and when few are laid for study as emphasis, then the economy creates pressure on them and the result is unemployment or even quacks. Consider for example: science Aerospace Engineering, Nuclear Engineering etc; Arts Law enforcements, communication radio and TV; Gerontology; Commercial Turf management, Business of banking rather than the much pressured Banking and Finance. Please, see Section

2 on the topic: courses in those scholarships, then try to compare our national level of competence to the United States, for instance.

The universities would not need to fix students except they applied first, to study courses as stated by their own records on the SCAP results and their attitudes interpretation.

PREDICTION 6: GOVERNMENT & LEADERSHIP

The issues of governance in every field is a very sensitive one. People tend to blame the government in power as much as they would blame the devil for every calamity in their lives including the results of their own negligence or mistakes. Agreed, the government ought to provide basic amenities and social infrastructure, but the citizens do also have a part to play. With respect to creativity and its philosophies of raydealism, I have done an x-ray on the relationship between the government and her own citizens and vice versa. I believe, the outlined principles would go a long way to achieve positive outcomes in our reasoning factor. Remember, you are in control of your mind by your choice…

These 18 relationships of creative governance are:

1. The government, as a group of persons, is the mirror and product of the mind of the citizens.
2. The people deserve the government they get because they allowed them in power.
3. The people who govern are the reflection of their own citizens.
4. If the citizenry of a nation is corrupt, it's most likely that a majority of people in the government would also be corrupt.
5. The character of a people is the index of the thoughts of their government as workable in their laws and perceptions.
6. A creative people would judge their government based on creative works; for nothing speaks volume like works.
7. Since creativity individually is service and resourcefulness, then those who desire to govern would consider it an opportunity for greater service.
8. A creative economy is a productive economy and thereby progressive.

9. A nation is best judged by the continual reduction of poverty, and the creation and implementation of policies to create jobs and opportunities.
10. A nation's resources must be creatively developed if its overall income is to be redoubled.
11. Corruption exists greatly in an economy when it is overtly dependent on other economies.
12. Corruption in an economy is an index of "Survival fear", that is fear of the future or their survival, because they are less creative, more dependent, less stable.
13. When people are creatively prosperous at the individual level, they tend to grow to a point of personal stability and freedom where they are begged, urged or appealed upon to leave their works and consider governance. Here, it is considered a waste of resources and creative time to go and sit in government rather than increasing their worth by their individual works.
14. Since creativity demands transparency of work, great leaders would be considered people who were transparent in leadership and governance.
15. In a creative economy, the statistics would prove a trend and thereby become a guide to the government on what infrastructures and policies to adopt in order to further productiveness in the nation.
16. Creativity, to be effective nationally, must be based on provable records, statistical trends and less dependence on government for livelihood.
17. The progress of the individual based on creativity encourages others, inspires confidence and leads to a progressive economy; sustaining less dependence, and more independence and individualism.
18. In creativity, a greater population would mark a greater market for the distribution of such unique product or service thereby reducing greed and the fear of population explosion.

The dynamic nature of the universe is a clear testimony to creativity; there's always a better way to do things and the earlier we inculcate this course in our study curriculum in preparation of the minds of youths to face the challenges of the emerging age, and globalization, the better for us. This would ensure we transit successfully from the Information / Jet age to the era I describe as the Info-Time age; the creative age formally a process where age is not a determinant of greatness; since the youths would likely begin earlier to process their minds on time. Nevertheless, creativity development is a collective responsibility. Let competence not wealth prove your worth.

CHAPTER 50

WORKING & LEADING?
OR
SLEEPING & LOSING?

My last advice to the aspiring creator would end in a series of questions, which I have derived from the two most determining questions stated above.

Firstly, are you a reader? If no, you need to develop this habit for inspiration could come from reading. If yes, what kind of stuff do you read the most. According to Charlie Jones, "everything in your life will likely remain the same in the next five years except the friends you keep and the books your read". You need to read materials which will develop your mind and increase your self worth. Reading is a better way of obtaining strategies within a limited time while others struggle day and night to find the wining solution. Reading is part of effective work and working strategically keeps you leading. A leader is made from within and not the day the world cheers him. No one who is not sure of himself can get involved in a competition and if he does, he does not hope to win a prize which he very well knows that he does not deserve. While working, what motivates you? is it the pay pocket or the gold? Is it greed or excellence? Is it innovation or piracy? What are those goals you need to set? You need to constantly re-evaluate your working philosophy and guiding principles.

Secondly, are you still sleeping? I have never seen a man who received honours for long sleep; for the longest sleepers are the dead. Do you still snooze so many honours of your life away? The sleep of a creator is a pleasure for him since it is the necessary rest; but an emblem for laziness for the coward whose body, though not finding a job to do, fails to search his mind and put the cells of his sticky brain to good use. Sleepers, I shall say again and again, are losers; the sleeping head gives no faithful account of his time and life to his creator, nor to the world in which he has been born.

The mystery of sleep is simply the fact that it is never enough, never satisfying and always demanding.

Citizens of my dear existence, please do consider the actions you take on a daily stance. In your present mood, time gives you no hoot nor cares about your livelihood for to it, you owe your life account. Always ask yourself honest questions and giver yourself honest answers. Today, it is important that you know whether you are working and leading or sleeping and losing. You, yourself alone, are the first judge to administer judgment upon your deeds…

Thank you

Your Excellency, yours sincerely,
My dear Student Me.

ASPIRATIONS OF THE CREATIVE AGE

By the Grace, Power and Mercy of the Almighty God, the creator of the Universe and man, ultimately in His own Image and likeness; we solemnly believe and work to fulfill its stated aspirations as follows:

i. In the Creative Age:- There shall be more jobs available than the applicants;

ii. In the Creative Age: People shall not be measured by their level of affluence nor wealth; but by their positive and creative contributions to their society and humanity.

iii. In the Creative Age: Governance shall be transparent, reflecting itself as the product of the thinking of the people; and leaders shall be judged not upon the prejudice of race, colour, ethnicity, or bigotry but by the records of contributions to human dignity and service to God.

Greatness, Greatness and nothing less!

Thanks be to God.

Reference Glossary on Pmakumatics

Expressed as diagrams, formulae and deductions

1.

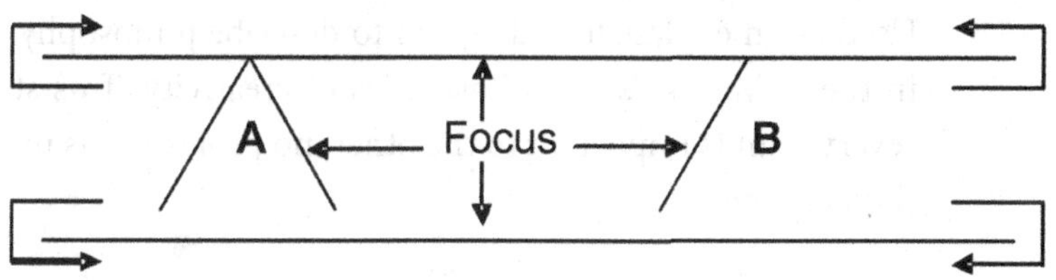

Diagram Title: POINT 1: UNIVERSAL RELATIONSHIPS
This is the first proof of creative bonding found in creative psychology.

2.

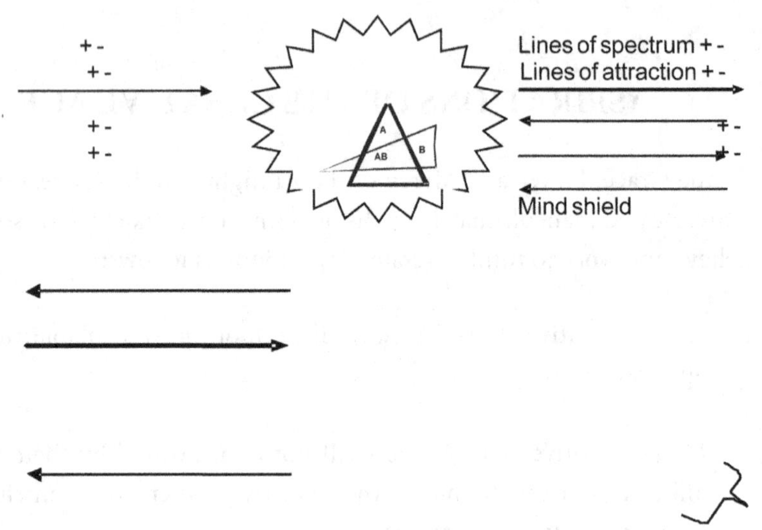

Diagram Title: <u>POINT 2: ELECTRICAL AFFINITY</u>
This is the second proof of creative bonding found in creative psychology

3.

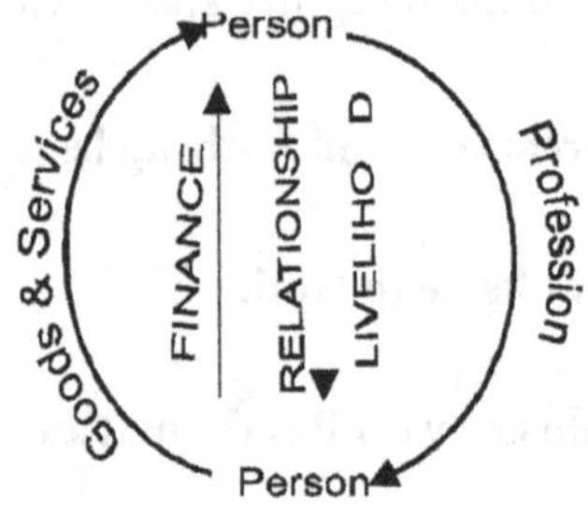

Diagram Title: <u>CYCLE OF PROFESSIONS</u>
Used as an explanatory diagram to describe philosophy 9, found in the P. M. Aku's 60 philosophies of creativity. This states that "every man is important to the other; no profession is more noble than another".

4. Interest $= \dfrac{CU \times TM \times PI}{OCC}$

$= \dfrac{CU \times TM \times PI}{OCC}$ = In,

Where Curiosity = CU
Time = TM
Place = PI
OC = Occurrence

$\dfrac{CTP}{OC} = I$

5. Interest, exists in different levels known as Thermometric Thinking Levels (TTL or T^2L).
These are: (1) Low TTL (2) Medium TTL (3) High TTL

7. The categories of TTL are stated below:

High TTL	=	75% - 100%
Medium TTL	=	45% - 75%
Low (Grade II)	=	15% - 45%
Low (Grade I)	=	0% - 15%

8. P. M. Aku's law of knowledge:

$\dfrac{Kn \quad Ex}{De \,\&\, Ab} = TP \text{ or } KTP = \dfrac{Kn \quad Ex}{De \,\&\, Ab}$

Where KTP = Constancy of Time (T) and Place (P)
= Sign of proportionality
De = Decision
Ab = Ability

9.

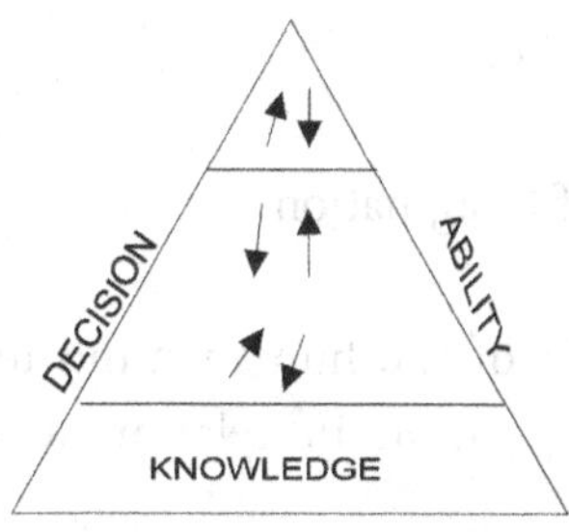

Peak of Aspiration

Ascending or Descending levels of knowledge.

Diagram Title: Triangle of knowledge. Used to express further in line with the formula, the process of knowledge in ascending and descending levels inherent in the mind of the creator.

10. $\frac{\text{C.P.}}{\text{CME}} = \text{H}$

A formula used for the expression of the graphical analysis of the relationship between the creative mind energy (CME) and creative passion (C.P).

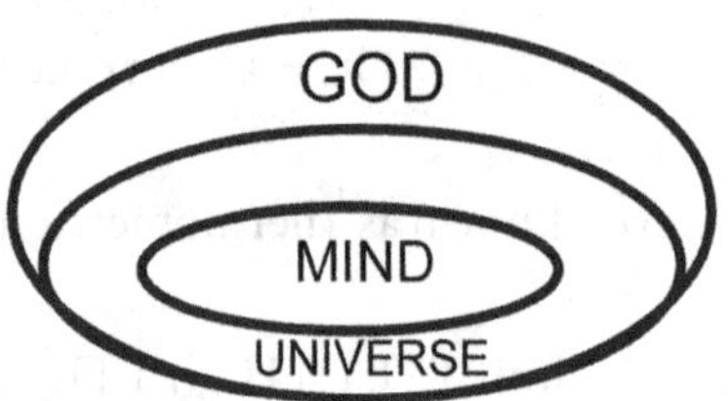

11.

Diagram Title: Central Universal Process (C.U.P)
Used to describe the central mechanism of universal control endowed by God to man. Found in mind mechanics and engineering.

12. Level of endowment

$= \frac{\text{Total Pointing Arrows}}{\text{Average Point}} \times \text{Peak of Aspiration}$

$= \text{LOE} = \frac{\text{TPA}}{\text{AP}} \times \text{POA};$ or better still

$\text{LOE } \frac{\text{TPA}}{100\%} \times 1{,}000\% \text{ IE}$

Used to calculate the level of endowment in order to determine the quantity of potentials in a person's ability.

13. $\frac{\text{Imagination}}{\text{Desire}} \times 1{,}000\% \text{ IE},$ or

$\frac{\text{Im}}{\text{De}} \times 1{,}000\% \text{ IE} =$ Law of Imagination

Used to prove the relationship of the human mind to desire, imagination and peak of Aspiration in relation to creative success.

14. Purpose may be defined or expressed Pmakumatically as

$$\frac{I.\ P. + D.\ P}{IMAGINATION} \quad E.P;$$

Where I. P. stands for Initial Purpose
D. P. stands for Defined Purpose
E. P. stands for Established Purpose

15. $$\text{Education} = \frac{\text{Educator (ET)} + \text{Acceptor (AT)}}{\text{Time (T)} \times \text{Place (P)}}$$

$$E = \frac{ET + AT}{KTP}$$ Where K is a sign of constancy, because Time and Place defines our environment

This is used for educational analysis and found in creative education

16. $$\text{Learning} = \frac{\text{Interest} \times \text{Attention}}{\text{Ability}} + \text{Time} + \text{Place}$$

$$L = \frac{In \times Att + KTP}{Ab}$$ Where K is a sign of constancy

This is used to define real or true learning processes in a person. Each of these constituents must be indicated to a maximum for proper results to occur in creativity and educational functionality.

17. peak of Aspiration [POA] = 1.000% IE

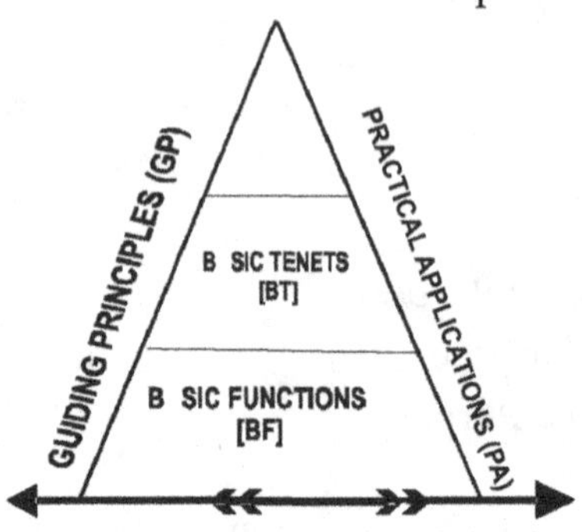

Line of Reciprocals (LOR)
Diagram Title: Learning Constituents Triangle (LCT)

The LCT is used to express diagrammatically the major constituents of learning or moreso, proper learning.

18. Learning constituents is expressed in a Pmakumatical relationship as: Learning Constituents = $\frac{\text{Guiding principles}}{\text{Practical Applications}}$ x Peak of Aspiration

i.e. $\frac{\text{G. P.}}{\text{P. A.}}$ x 1,000% IE

Where 1,000% IE is used because there is no limit to learning and its various aspirations;
IE stands for Infinitum Endowment which means limitlessness to every natural endowment.

19. The instruments of creative sciences i.e. the Pmakumeter Guage, Diary of ideas, creativity analysis sheet; pictorial (Calender type) representations of the structure of the creative and parts of mind Dichotomy, the plastic models of the structures of the creative mind and mind Dichotomy etc, form the prescribed instruments in Pmakumatics, which are basic functional parts of the laboratory or theatre of the study of creative sciences and the professional creative scientist. Other instruments e.g. Pen, pencil, measuring rules or tapes, painting materials etc which are used for specific lines of creating or inventing are called general resources. They are specific to the aspect of creativity which a person decides to make an endeavour.

FAST FACTS ABOUT CREATIVITY

1. Through the ages, I have seen a force grow up slowly, I have sensed its power upon every one who welcomed it, its hurts against everyone who resisted it, and have concluded one thing: creativity is you, you are creativity, if you welcome it, you open your opportunities yet if you resist it; your own wounds would never heal.
2. The connecting thread between you and your creator is the creative instinct; to see, to solve and to win; then to reach the creative prowess.
3. Yes, we do need revolutions, or perhaps, a revolution in our governments and the economy, but today I tell you, its not violence, for this results in mess; it's not religious piety, for God expects results from your working sweats, your thinking cap; it's not moral cowardice born out of pretence and many failures; it's mental revolution, original ideas for moral re-armament; an intellectual revolution for sincere change and dynamism.
4. Creativity is called the silent Empire because it is a fire burning within the person, the results come to view to take its place among the system of things.
5. It's a crime, I foresee, in few years to search for the job you didn't create pride and false assumptions have stupified our great ability up to this date.
6. Creativity is not an institution; it's not a point of view; Creativity is mental revolution such that starts in you.
7. The common wealth of heaven way be reached by the heart; the common wealth of the earth is created by the creative mind.
8. Man, over the ages has proved to be egocentric, selfish and stupid; he has shared every thing except that which he was originally created to share; his creative ability in all docility.
9. Everyone must be developed and given a chance, a right to express his creative ability, that is one last right implied but not expressly stated: the right to creative expression, and its limits; for human dignity.

10. Do you recall the words of Dr. Bucham; "Before a God-led unity, every problem will be solved. Empty hands will be filled with work, empty stomachs with food and empty hearts with an idea that really satisfies". But I say, only real creative ideas satisfy these records.
11. The creative mind has no time for sorrows, nor complaints, nor failures of stupid gossips and idiosyncracies, he is simply solving problems he has found.
12. All over Europe, the Americas, Asia, Africa and beyond, our races may differ in colour, but our blood is still red and our brains structured to be bred.
13. Whether the European giants, the Americas super powers, the Asian tigers, the African strengths; one thing common to them is problems and the ability to solve them.
14. Creativity is the foundation of national life, as it is of the very foundations of the universal existence.
15. Creativity is the God's source of hope to mankind, with the real purpose practical application to the standards of integrity, for by listening, God still expresses this love in His variety through us.
16. Men must learn to have a faith that will create the right revolution think through my mind and have such faith; see through my acts and make such change for I have learnt it and felt it, now it's your turn to experience it.
17. The creative mind is a moving industry a moving library, a moving ocean; such that if everyone manifested his, then the world would need thrice its present population to meet with its demands for labour.
18. Let it be known today that working for someone else ought to be to gain experience, more it is a crime when beyond ten years for that space you fill hinders another's chance; you ought to be starting something else.
19. Life is a market of ideas, a banquet of delicacies, men are hungry for living and knowledge creative endowments is man's own privilege.
20. Drop your endowments, you're become mere ornaments; imitate others and you lose your order. Believe in God, in everything else and your creative process, it does not fail and has never failed.
21. Creativity is the mustard seed, small enough to bring liberation big enough to encompass the world worthy to die for; great to live in.

22. Every time we respect nature, there's no mental torture in order to be perfect, we must remain natural.
23. Now I understand the sayings: "I looked at those leaves, beautiful and yet so near to their end, and I felt a deep longing to find an idea big enough to free humanity from chaos and confusion".
24. Now I realize that the greatest slavery is not financial but mental, fear and despondency; one full of creative droughts.
25. If the study of creativity sets men free then I think, Nigeria, Africa may have made the greatest of all findings the best contribution to the world society, and perhaps, I never to be forgotten; yet there lies a lot more hidden to be seen.
26. Creativity speaks many languages; it depends on which one you understand; many professions, it depends on which one or two you wish to practice.
27. The Great men of the ancestral pedigree agree that there is a power within they lived with it, and have experienced it; but they may not have known it is creativity.
28. Creativity is a power, you exercise it, it is a living force, you experience it, it is a driving passion, you can feel it, it is a thinking tank, a mental sharpener, it's no mythical legend, but a living reality you see it, fee it, work it and enjoy it; the connecting link, it makes all the difference.
29. The greatest tragedy of life is to pretend to be what you're not; living the life of someone else.
30. Clear those mental cobwebs, creative endowments keep you ahead of all life's taxes and payments, yours is free of all universal elements.
31. The essence of humanity is creativity. Creativity is the mother of all education; all encompassing and its functionality is collective, either in the Arts, Sciences or the Business terrain of studies and practice.
32. There is no proper education achieved in the past or the present civilization until the creative endowments, talents and potential of every individual is discovered, measurable and developed purposefully.
33. The greatest of human discovery is the power of creativity endowed in the. human person, the worst of it is in its misconception and misuse; while the loftiest of its achievement lies at the structuring and integration of its study into the present and future educational system.

34. What should bother us is not that great inventions were made as in works of creativity achieved by few men and women; but what really happened within those people that led to their greatness.
35. Blame the government as much as you blame yourself, you are a moving bundle of opportunities, the government needs more direction from your wealth of productive ideas, precepts, experiences and examples.

MEET PETER MATTHEWS AKUKALIA CH. S.: CREATIVE FACTS ABOUT THE WORLD'S 1ST CREATIVE SCIENTIST [WFCS]

By this work and others, Peter Matthews Akukalia is:

1. 1st person in history to develop a field of study single handedly.
2. 1st Chartered Sage (Ch.S) ever equivalent to a Double Star Ph.D.
3. 1st person ever to develop completely a form of calculations without a prior source and named after him.
4. 1st person ever to create an instrument to calculate the internal factors of the mind
5. 1st person to develop the complete interpretation of a person, thereby eliminating the one-on-one interview process inherent in organizations.
6. 1st person to develop the new kind of study called Psycho-social Science
7. 1st person to develop and prove the encompassing new philosophy called raydealism and the first raydealist ever.
8. Father of the 4th genre of Literature called Nidrapoe and the 1st drapoet ever
9. Father of the 5th genre of literature called Propoplay / Peterwrit, and the 1st Propoplaywright / Peterwrit.
10. Father of the "Silent Empire" the emerging creative age comprising the economy and governments of every nation.
11. He is the father of the proposed ministry of creativity of nations; the father of the Institute of Chartered Creative Scientists (ICCS); father of the Hall of Creators; and the author of the Educational Development Master Plan (EDEMP) and Creative Age Project (CAP).
12. By this work on the authoritative book: The Oracle; Principles of Creative Sciences, he has emerged as the World's 1st Creative Scientist (WFCS).

Opportunities / Benefits

1. Training and employment opportunities are large for the private sector and individuals.
2. Partnerships entitle governments and citizens to tap, from these resources.
3. Partnership with governments (Federal & States) PCS (induction to universities (to train strategists); SCAP (sec sch. High sch, JSS 3 & above)
4. By the principle of operation, creative scientists must be placed strategically in every sector of the economy; according to profession e.g. a creative accountant; a creative lawyer etc.
5. Professional creative scientists work at their level as required e.g. M;Cs, G;Sc, S;Cs, Ch.S etc
6. Creativity is the new mentality; it is the intellectual and economic revolution of the new age; the Creative Age.

Seminars organized periodically on the benefits of the Creative Age.

FROM THE AMBASSADOR

Principles of Creative Science (P.C.S) is not just a science text book or a book in science fiction, it is much more than that. This book is fundamental in every sphere of learning: Arts, Science, Psychology, Literature, Philosophy, Law, Commerce, Culture and Religion. It expounds creativity and Education in every area of study.

This book discusses the potentials and abilities endowed in humanity and how it could be analyzed, quantified and harnessed. A book by Dr. Wesley Duewel was called "Touch the World through Prayer". Mr. Matthew Akukalia is set to touch and revolutionalize this nation and the world academically by this work.

This book is an eye opener to further research works and creativity. No person will ever read and study this book and remain unemployed. The Oracle is the long awaited solution to the problems of unemployment and unfulfilment cutting across every aspect of humanity. Believe strongly that with time, Mr. Matthew Akukalia will become an ideology by his convictions.

This book should be recommended for undergraduates, graduates, post-graduates and researchers. By this work the Creative Age is born.
Cosmos Azubuike Micheals (Ajuonumah) Ambassador to the World's First Creative Scientist.

ABOUT THE BOOK

- Teaches how to discover and develop your creative potential
- Discusses the encompassing scope of creativity in all professions such as creative psychology, physiology and medicine; etc and the new genres of literature.
- Contains the philosophies, facts and myths, direct and experimental observations of creative endowments.
- Authoritative in mind mechanics and engineering, creative doctrines and Pmakumatics; Laws and guiding principles of statistical creativity etc
- Written to demystify the subject of creativity;
- First time ever by the esteemed African Czar, African Shakespeare, the 1st Chartered Sage and the World's 1st Creative Scientist himself.
- Projected to lead the "Silent Empire" the emerging, Creative Age, Creative Economy and Info Time Age.

SPECIAL CONSULT BOOKS

These books are included but do not form the exhaustive list of references, consulted, read or studied on the above discussed subject title.

They include:

1. Beyond freedom and Dignity by B. F. Skinner
2. The third Wave by Alvin Toffler
3. Life and Death of Adolf Hitler by Robert Payne
4. Current Biography Year Book by the H. W. Wilson company 1987
5. current biography Year Book by the H. W. Wilson company 1964
6. Math connections by William P. Berlinghoff; Clifford Sloyer; Robert W. Hayden
7. A Roman centurion by Stewart Ross
8. America Is; by Henry N. Drewry; Thomas H. O' Connor
9. The Holy Bible King James Version
10. New World Translation of the Holy Scriptures published by Watch Tower Bible and Tract Society of New York, Inc.
11. The Great Barrier Reef by Craig McGregor and the editors of Time Life books.
12. The Israelites (the emergence of Man) by the editors of time Life books
13. Lunar photographs from Apollos 8,10 and 11 published by the National Aeronautics and space Administration.
14. Alien by Paul Scanlon and Michael Gross
15. Iceland: life and Nature on a North Atlantic Island published by Iceland Review.
16. Times Past by R. J. Unstead and illustrated by Ivan Lapper.
17. Above and beyond: The Encyclopedia of Aviation and space Sciences Vol. 9, published by New Horizons publishers, Inc.
18. Great American Journeys published by Reader's Digest
19. Wildlife on your Doorstep: The Living Country side published by Reader's Digest
20. Nuclear Power by Nigel Hawkes; published by Gloucester press.

21. The Children's Books of Cars, Trains, Boats & plains by Kenneth Allen.
22. Career Choices: A guide to making the right decision by Judith Midgley Carver & Amanda Duckett.
23. The Children's bible in colour; published by the Hamlyn Publishing Group Limited.
24. The American Heritage Dictionary: Third Edition; published by Dell publishing.
25. Collins Gem English Dictionary: An imprint of Harper Collins publishers.
26. The Secret of family happiness published by Watch Tower bible and Tract society of New York, inc.
27. Love notes to our moms and other women of influence published by speaking of women's Health.
28. Reasoning from the scriptures published by Watch Tower Bible and tract Society of New York, Inc.
29. The Power of positive Thinking by Norman Vincent Peale.
30. How Never to be tired by Marie Beynon Ray
31. Jehovah's Witnesses: proclaimers of God's Kingdom published Watch Tower Bible and Tract Society of Pennsylvania.
32. My Book of bible stories; published by Watch Tower Bible and Tract society of Pennsylvania.
33. Benefit from Theocratic ministry school education ;published by Watch Tower bible and Tract society of New York,Inc.
34. All scripture is Inspired of God and beneficial; published by Watch Tower Bible and Tract society of New York,Inc.
35. Revelation: Its Grand Climax At Hand! Published by Watch Tower Bible and Tract society of New York,Inc.
36. The best of France: A cook Book; published by Collins publishers,San Francisco and produced by smallwood and stewart,Inc.New York.
37. Learn from the great Teacher; published by WatchTower bible and Tract society of Pennsylvania.
38. The businessman's pocket book by Efeoghene Okposio
39. Goodnews for people of all Nations ;published by Watch Tower bible and tract society of Pennsylvania.